"*How to Be a Great Cop* is a must reading for every police officer in this nation, as well as any contemplating a career in law enforcement. It provides an honest, straightforward, and total insight of what police officers can do to achieve excellence."

Ed Nowicki
Founding Executive Director
American Society of Law
Enforcement Trainers

OTHER BOOKS BY NEAL TRAUTMAN

The Cutting Edge of Police Integrity
How to Protect Your Kids from Internet Predators
Law Enforcement Training
Law Enforcement In-Service Training Programs
Law Enforcement-The Making of a Profession
A Study of Law Enforcement
Landing a Law Enforcement Job

How to be a Great Cop

Neal E. Trautman

Prentice Hall
Upper Saddle River, NJ 07458

Library of Congress Cataloging-in-Publication Data
Trautman, Neal E.
How to be a great cop/by Neal E. Trautman.
p. cm.
Includes index.
ISBN 0-13-032474-4
1. Police--Vocational guidance--United States. 2. Law enforcement--Vocational guidance--United States. I. Title.

HV8143. T725 2002
363.2'023'73--dc21

00-059841

Publisher: Jeff Johnston
Senior Acquisitions Editor: Kim Davies
Production Editor: Linda Cupp
Production Liaison: Barbara Marttine Cappuccio
Director of Manufacturing and Production: Bruce Johnson
Managing Editor: Mary Carnis
Manufacturing Manager: Cathleen Petersen
Art Director: Marianne Frasco
Cover Design Coordinator: Miguel Ortiz
Cover Designer: Bruce Kenselaar
Cover Image: © Tracey L. Williams / Courtesy of the Bernards Township Police Department, Basking Ridge, New Jersey.
Marketing Manager: Ramona Sherman
Editorial Assistant: Sarah Holle
Interior Design, and Composition: Lithokraft II
Printing and Binding: R. R. Donnelley & Sons

Prentice-Hall International (UK) Limited, *London*
Prentice-Hall of Australia Pty. Limited, *Sydney*
Prentice-Hall Canada Inc., *Toronto*
Prentice-Hall Hispanoamericana, S.A., *Mexico*
Prentice-Hall of India Private Limited, *New Delhi*
Prentice-Hall of Japan, Inc., *Tokyo*
Prentice-Hall Singapore Pte. Ltd.
Editoria Prentice-Hall do Brasil, Ltda., *Rio de Janeiro*

10 9 8 7 6 5 4 3 2 1
ISBN 0-13-032474-4

Dedicated to my daughter

Heather Rene Trautman

The author wishes to acknowledge
sincere gratitude
to his wife, Robin,
for her untiring effort
and endless love.

Contents

FOREWORD

"It takes a special kind of person to be a police officer. They must remain alert during hours of monotonous patrol, yet react quickly when need be, switch instantly from a state of near somnambulism to an adrenaline-filled struggle for survival, learn their patrol area so well they can recognize what's out of the ordinary.

It takes initiative, effective judgment and imagination in coping with a complex situation, family disturbances, potential suicide, robbery in progress, gory accident, or natural disaster. Officers must be able to size up a situation instantly and react properly, perhaps with a life or death decision.

Officers need the initiative to perform their functions when their supervisor is miles away, yet they must be able to be part of a strike force team under the direct command of a superior. They must take charge in chaotic situations, yet avoid alienating those involved.

They must be able to identify, single out and placate an agitator trying to precipitate a riot.

They must have curiosity tempered with tact, be skilled in questioning a traumatized victim or a suspected perpetrator. They must be brave enough to face an armed criminal, yet tender enough to help a woman deliver a baby. They must maintain a balanced perspective in the face of constant exposure to the worst side of human nature, yet be objective in dealing with special interest groups. And if that isn't enough, officers must be adept in a variety of psychomotor skills: operating a vehicle in normal and emergency situations; firing weapons accurately in adverse conditions; and strength in applying techniques to defend themselves while apprehending a suspect with a minimum of force.

Then, when it's all over, they must be able to explain what happened—in writing—to someone who wasn't there in such a way there's no opportunity for misunderstanding and to document their actions so they can relate their reasons years later."

Bill Clede
Retired Officer and Trainer
Member, American Society of Law Enforcement Trainers
Charter member, International Association of Law Enforcement Firearms Instructors
Technical Editor, *Law and Order Magazine*
Author of Police textbook

PREFACE

Law enforcement—it is one of America's greatest, yet most difficult, professions.

A heritage of pride. A history of tradition.

This book is for people who thrive on overcoming obstacles and working hard for a good cause. Whether you have yet to be sworn in, are on the street right now, or have become a seasoned cop; you should have a thirst for knowledge. *How to Be a Great Cop* shares the insights and experiences of those who have "been there." It talks about reality and the things that are important in an officer's life.

Chapter 1 focuses on the history of law enforcement. Everyone benefits from the insight that only experience can deliver. When you understand police tradition, you'll better understand yourself, your fellow officers, and the future of policing.

Chapter 2 is about greed, honesty, temptation, and pride and the choice that must be made between them. It's about ethics and honor. It doesn't preach, but talks straight about things that may ruin your career and life.

Chapter 3 responds to the question of what it takes to be a great cop. Highlights include personal control; discipline; wisdom; caring; positive attitude; hard work; supporting others; and staying educated. You have to work hard to have the "right stuff."

Chapter 4 discusses interacting with the judicial system. Some of the subjects reviewed include the law, American justice, prosecution, courts, civil suits, plea bargaining, defense attorneys, and legal terms. If you learn how to make the most of the legal frustrations with which our criminal justice system confronts you, you will have enough knowledge to do your job and save yourself a lot of grief.

Chapter 5 discusses surviving the street, including strategies and techniques of the nation's leading survival trainers. Mental conditioning is emphasized. Practical, realistic techniques can make a difference when you need them. The chapter provides important information on anticipating danger, understanding physiological changes in a crisis, and analyzing shoot outs.

Emotional survival is often overlooked in police work. Officers suffer serious medical problems or die from stress-related difficulties more often than attacks from guns or knives. Chapter 6 sheds light on these issues. It examines why people become cops, the natural changes in emotions anyone can expect, and how to manage police stress. In addition, it offers information on dealing with angry citizens, your spouse, and children; the truth about exercise; and special concerns of veteran officers.

Chapter 7 discusses why it is important for an officer to be physically fit. It presents helpful information on physical exams, stretching, choosing the right exercises, nutrition, and eating habits.

Appendices A-D provide information on law enforcement agencies and training organizations, related associations, and state training councils and police standards commissions. Contact information for more than two hundred important organizations are included. Appendix E outlines the educational requirements of law enforcement agencies across the country. It is part of a Bureau of Justice Statistics report entitled *Law Enforcement Management and Administrative Statistics, 1997* (NCJ 171681). Understanding the facts, figures, and statistics about the operations of contemporary law enforcement is a priceless tool for any officer to achieve greatness.

Every chapter has been written to cover specific subjects with reasonable thoroughness. The underlying premise is that most people have the ability to be successful. The extent of their commitment determines whether they will achieve greatness. To be a great cop, attitude and commitment are everything.

ACKNOWLEDGMENTS

The author acknowledges his appreciation to the following law enforcement authorities for their thoughtful comments:

Bill Clede	Law Enforcement Author
Timothy M. Dees	Law Enforcement Trainer and Author
Louis M. Dekmar	Chief, La Grange, Georgia, Police Department
Lee D. Donohue, Sr.	Chief of Police, Honolulu Police Department
Corinne Garrett	Lieutenant, Orange County, Florida, Sheriff's Department
Paul Harvey	National Media Commentator
John Linn, Ph.D.	Criminal Justice Department, Altoona College, Penn State University
Peter C. Loomis	Professional Standards Coordinator, Maitland, Florida, Police Department
J. Dale Mann	Georgia Law Enforcement Academy
William A. May, Jr., Ph.D.	Major, Louisville Fire/Police Department, Louisville, Kentucky
Ed Nowicki	Former Executive Director, American Society of Law Enforcement Trainers
Bernard C. Parks	Chief of Police, Los Angeles Police Department
Richard Pennington	Chief of Police, New Orleans Police Department
Larry Plott	Former Director, Peace Officer Standards and Training Commission, Idaho
Michael Sanford	Administrative Aide to Chief of Police, Seattle, Washington
J. E. Tillman	Chief, North Las Vegas Police Department
August Vollmer	"Dean" of Modern Policing

ORGANIZATIONS

Federal Bureau of Investigation
National Institute of Ethics
Police Executive Research Forum, International Association of Chiefs of Police (IACP)
International Association of Directors of Law Enforcement Standards and Training (IADLEST)
U.S. Department of Justice, Bureau of Justice Statistics
World Rights Watch

Abraham Lincoln was raised in poverty.

Franklin Roosevelt was brutally struck down with childhood polio.

Elementary school teachers told Albert Einstein he was "retarded."

Glenn Cunningham, who broke the world record in the one-mile run, was told by doctors when he was a child that he would never walk again.

The most important aspect of courage is *never* giving up . . . when it's important.

"I shall not pass through this world but once. Any good, therefore, that I can do let me do it now. Let me not defer or neglect it, for I shall not pass this way again."

Unknown Author

"It is my belief what it takes to be a great cop is:

When sworn into office, the sacred Oath is taken, is held to its highest regard through their entire law enforcement career and even beyond. The badge worn is not just a piece of metal, but a symbol of the public's trust that will not be compromised, under any circumstances.

Officers must perform and adhere to the department's Core Beliefs of Integrity, Fairness and Service. Integrity, dedicated to maintaining the highest moral and ethical standards, through the principles of pride, honesty, trust and courage. Fairness, dedicated to treating citizens with dignity, respect and equality. Service, committed to developing a well-trained, highly motivated and courteous officer to serve the community and the organization with pride and professionalism.

Officers must possess the ability to forge partnerships with diverse communities for the purpose of building confidence and accountability to the public they serve.

They must continue their academic education for upward mobility in their organization and to keep abreast of the latest police policies, practices and techniques in modern law enforcement. Their knowledge of today's policing must go beyond a single jurisdiction, but expand into how other law enforcement agencies, throughout the United States and abroad, are progressing.

The job of today's law enforcement official is difficult because of the many responsibilities he or she are required to perform, daily, but be a "great police officer" is limited to just a few. They are the men and women in policing who have gone beyond the normal scope of their performance and truly exemplify the meaning of law enforcement."

Richard Pennington
Chief of Police
New Orleans Police Department

"Compassion. The really great cops that I know—the ones that I really respect—are very compassionate people. They really care about others. Sure, some seem to be pretty crusty on the outside after twenty years of watching man's inhumanity to his fellow man. But deep down inside, under that cover, there's a person who really cares. That's what keeps them going after twenty years. They still wear the badge and put on the uniform each day, going to work really believing that they're going to help at least one person today. At least one person today will be alive, or somehow be better off because of something that they did. That quality comes from within. It can't be taught at any academy or in-service class. It doesn't show up on any written or oral exam. But it's there. The cops that don't have it aren't as good as they could be. It's the one thing that separates the good cops from the great cops. A really great cop keeps that compassion for their entire career."

Bill May, Jr.
Captain
Louisville Fire/Police Department
Louisville, KY

"Keep away from people who try to belittle your ambitions. Small people always do that, but the really great make you feel that you, too, can become great."

Mark Twain

"My mother taught me very early to believe I could achieve any accomplishment I wanted to. The first was to walk without braces."

Wima Rudolph
Four-Time Olympic Gold Medalist

CHAPTER 1

COPS—THEN AND NOW

"A police officer is expected...
to have the wisdom of Job,
the kindness of the Good Samaritan,
the strategy of Alexander,
the faith of Daniel,
the diplomacy of Lincoln,
the tolerance of the Carpenter of Nazareth,
and finally, an intimate knowledge
of every brand of the natural, biological,
and social sciences.
If he had all of these he might be
A good policeman."

In loving memory of
the late August Vollmer
"Dean" of Modern Policing

Cops — so many things to so many people. To a crime victim they may be heroes. The next person may call them incompetent fools. A small child may look at them with admiration. A teenager may glare at them with arrogance or hatred.

Every experienced street cop has felt the frustration of having to deal with the same "low life" over and over. Yet they are expected to take it all in stride and be ready to handle the next call with wisdom, courage, and patience. Sometimes they make serious, yet honest, mistakes and few will be able to forgive themselves.

What type of person would choose to face daily animosity and ridicule? Who would want to work the long, miserable hours of midnight shift, stand in the rain or cold and direct traffic, or live with the discomfort of wearing a bulletproof vest during the summer? Who would want to live with the glares and snide remarks from motorists stopped for traffic violations? Veteran officers know and understand the peculiar kind of isolation that cops feel when walking uniformed into a restaurant or being identified as cops at a party.

Why do people become law enforcement officers? They usually are sincere, enthusiastic, and driven by good intentions, but, above all, they simply want to help others. These feelings will probably change to some degree as the years pass as enthusiasm and sincerity gradually mix with frustration or cynicism. Without good leadership, though, cops' dedication to duty can resist disintegrating into apathy and arrogance.

This book is dedicated to the principles for which thousands of officers have died. The highest ideals of law enforcement must become ingrained in all officers, so that those who die will not do so in vain. Loyalty, dedication, and integrity cannot be the brunt of jokes or be taken lightly—serving others is the foundation of good policing.

EVOLUTION OF POLICE

When aspiring military officers begin their careers in Annapolis, they spend considerable time studying the tradition and history of the American military. In this and many disciplines, the wisdom and insight acquired by understanding the past is invaluable. A deep sense of pride and tradition creates sincerity and understanding of purpose.

Police departments, unfortunately, rarely appreciate the value of such training programs. Pride, dedication, and comradeship could be developed by understanding the sacrifices made by former officers. Such an appreciation of the past could become the foundation for building the future.

Beginnings—Early 1900s

Prior to 1900 law enforcement left a lot to be desired. Some cities began to pass laws that replaced lazy and incompetent night watchmen with officers. Although this was a considerable improvement, police protection usually consisted of merely a single constable who worked during the day.

As cities grew, the number of officers increased accordingly. Yet graft (illegal or unfair gain) and corruption were widespread throughout the political arena, and political interference manipulated a chief's authority to appoint, assign, or dismiss officers. There were no employment standards. Fairness within the workplace was often nonexistent. Lack of employment procedures combined with political favoritism caused extremely low morale.

Lack of employment requirements also resulted in a relatively low salary. Officers wore no uniforms and the numbers on their copper badges were the only means of identification. Officers carried a 33-inch nightstick as their only weapon of defense. Still, political favoritism continued to flourish.[1]

During these early days, a few farsighted individuals realized that officers could become proficient with special training. One individual, August Vollmer, town marshall and later chief of police of Berkeley, California, knew that patrolmen must recognize criminal acts and the elements for proving convictions. He believed that officers must understand how and when to use force. Further, Vollmer understood the grave responsibilities placed upon law enforcement.

In the early 1900s many officers voiced their dissatisfaction with work conditions. The famous Boston Police Strike of September 9, 1919, originated from union settlement. By the turn of the century local police unionization had begun. The Boston union had requested permission to affiliate with the American Federation of Labor. The police commissioner refused and suspended several officers for their union activities. The union reacted by voting to strike. Thousands of officers walked off their assignments, sending Boston residents into panic.

After eight deaths, nearly a hundred injuries, and in excess of $1 million in property damage, President Wilson expressed the nation's sentiment. He stated, "A strike of policemen of a great city, leaving that city at the mercy of an army of thugs, is a crime against civilization. In my judgment the obligation of a policeman is as sacred and direct as the obligation of a soldier. He is a public servant, not a private employee, and the whole honor of the community is in his hands. He has no right to prefer any private advantage to public safety." [2]

Roaring 1920s

Police professionalism was struggling during the 1920s and 1930s, and social unrest was placing tremendous demands on law enforcement. In addition to the passage of the Eighteenth Amendment, which prohibited the manufacture, sale, and import of liquor, the Volstead Act of 1919 made provisions for enforcement of the Eighteenth Amendment.

Prohibition was repealed in 1933, but until then, it created a great deal of difficulty for the officers who were required to enforce it. The majority of citizens ignored prohibition laws, and the police became extremely unpopular. Graft, political influence, and corruption among the police force substantially increased because officers were frustrated.

A large percentage of the population frequently offered bribes to police officers. As prohibition continued to be unpopular, a general feeling of apathy toward police authority and the belief that officers could be "bought" resulted in extremely low morale and self-respect within police departments.

America's response to the turmoil, lawlessness, and corruption of the 1920s was the creation of a presidential commission commonly referred to as the Wickersham Commission. It scrutinized virtually every aspect of the criminal judicial system. An assortment of recommendations and suggestions were developed. Many recommendations were directed toward law enforcement management and operations.

Throughout history most law enforcement agencies have attempted to hire only high-caliber applicants. New recruits help create an agency's image. Unfortunately, America's law enforcement image suffered a setback as a result of manpower shortages during World War I. The lack of manpower resulted in unqualified and poorly skilled individuals entering the ranks.

Police recruitment was difficult in this era. Years later we still live with some of the personnel procedures created decades ago. Police personnel selection today lacks uniformity throughout the nation. Further, a definite relationship exists between the quality of officers hired and the quality of performance within an agency. If agencies fail to develop a thorough, effective hiring process, officers with character flaws will be hired. As a result, corruption is likely to occur when such officers face the temptations that a law enforcement career presents.

Many attempts have been made to develop consistent police professionalism. As an example, a civil service merit system was established to provide an avenue to escape political sponsorship and manipulation. This system rewarded merit and provided a means to deal appropriately with the problems of graft and corruption. However, experience has shown that civil service frequently protects the inept and lazy. It sometimes becomes counterproductive to professionalism by preventing the termination of those who should be terminated.

Growth and Struggle—1930s, 1940s, and 1950s

Higher education began to flourish in the 1930s. The University of Chicago offered police-related courses as part of its undergraduate curriculum. San Jose College began a two-year criminal justice associate degree in 1930. In 1932 J. Edgar Hoover continued his relentless pursuit of professionalism by developing the Law Enforcement Bulletin. Its initial purpose was to promote the advancement of police service through professionalism. To this day, it continues to offer realistic and practical assistance.

The International Association of Chiefs of Police created a safety division within Northwestern University during 1935. The division provided field services, promoted traffic safety, and conducted an assortment of valuable research. During the same year, Michigan State University began to require courses such as physics and chemistry in its police curriculum.

The Federal Bureau of Investigation's National Academy was established in 1935. The academy has played a crucial role in upgrading police service. It has trained thousands of administrative and supervisory officers for all levels of law enforcement.

The 1940s and early 1950s was an era of only mediocre professional advancement. The International Association of Chiefs of Police conducted a survey which indicated that officers throughout the nation were dissatisfied with their working conditions. Unsatisfactory working hours, salary, pension, and other employment benefits resulted in the formation and operation of 44 police unions by 1956.[3]

The quality of any law enforcement agency is dependent upon the quality of its personnel. Likewise, all cops want to have partners they can count on. Partners grow close together and often become good friends. One does not have to be a chief to understand that the most capable rookies will probably develop into reliable, confident officers.

Revolution—1960s

Unlike the 40s and 50s, the 1960s was a time when municipal police forces were in a tide of criticism. Student unrest exploded in the streets as our nation's colleges protested social conditions and the war in Vietnam. Civil rights demonstrations became violent. Riots were common in some major cities. The crime rate skyrocketed. Drug addiction climbed to unprecedented heights.

During the 1960s, law enforcement across the country was ill-equipped, unprepared, and poorly trained. The population had increased approximately 13 percent, and reported crimes had risen 148 percent—the front line of defense was struggling to enforce the law and maintain peace. Serious crime increased at a staggering rate: aggravated assault rose 102 percent, robbery climbed 177 percent, and rape climbed 116 percent during the decade.[4]

In 1964 J. Edgar Hoover stated, "More states need to make available police training. More universities and colleges should be initiating and increasing courses of study oriented toward the development of a police career profession. Law enforcement must raise its sights, broaden its outlook, and insist on a higher caliber of performance." Yet during the decade higher education for police was sporadic, to say the least. Fewer than 25 universities and colleges were offering any form of full-time law enforcement program.

The President's Commission on Law Enforcement, released in 1967, was a two-year extensive study on law enforcement by more than 2,450 staff members, advisors, and consultants. It was a far-reaching collection of reports, surveys, and statistical information, which is still a valuable resource today. Although the educational recommendations of the President's Commission have yet to be reached, efforts to establish higher levels of employment should continue.

The International City Management Association's municipal yearbook for 1968 indicated how badly police training was needed during the 1960s. The yearbook reported that police recruits received no training in 7 percent of all central city agencies, no training in 11 percent of suburban agencies, and no training in 32 percent of independent city departments. These statistics were for municipalities of 10,000 persons or more. In smaller cities less than 50 percent of new recruits received any police basic training.[5]

No states required formal police basic training until 1959. During the 1960s the International Association of Chiefs of Police conducted a study, which determined that the average police officer in America received less than 200 hours of formal basic training. Although the number of hours increased dramatically during the late 60s and early 70s, it is still disturbing, especially since so many other careers require more extensive training. Cosmetologists and barbers, for example, require several thousand hours of training. Embalmers receive more than 5,000 hours. Teachers must have in excess of 7,000. Lawyers are required to have 9,000 hours. Physicians receive more than 11,000 hours of training.

Why do these other endeavors require so much more formal training than law enforcement? It is because these fields appreciate the fact that poorly trained personnel result in poor performance. If accreditation boards and states did not require members of their professions to meet high standards, their efforts would not be recognized as professional. By establishing high training standards and rigid licensing for certification, high-quality service is ensured. All facets of policing—having far more serious responsibilities than virtually all other endeavors—must make the same solemn commitment to excellence.

Professionalism—The 1970s

"Police professionalism" became synonymous with higher education during the 1970s. Formal college degrees brought recognition to both individual officers and their agencies. Still, the value of education was a subject of considerable debate, as many veteran officers without a college education felt that the only education of any value was learned on the street.

The 70s became a time of awareness for most agencies, and officers were subjected to an assortment of new perspectives of society. As many officers continued their education, new viewpoints and a deeper appreciation for the role of the police were developed. It became apparent that all time-honored professions demand high standards of qualification and performance. A variety of courses related to problems within the criminal justice system showed that if the police were to be considered professionals, they must comply with high standards of performance.

Officers were required to comply with many strict legal requirements and limitations. Social problems plagued the police. Crimes related to drugs and economic conditions continued. Traffic difficulties worsened.

Officers who began their careers in the early 80s recall veteran officers stating that the job was different that it was in the 70s. The veteran officers usually said it was not as much fun being a cop as it used to be. They spoke of legal restrictions that kept them from doing their job. Some resented what they referred to as "a new breed" of cop. Not only was their job changing, fellow officers were becoming more educated and had different views.

Society now demanded more of the police. Officers were required to handle more complicated human behavior and be armed with more and better defenses. They appreciated the need to be equipped with physical and intellectual abilities for regulating, directing, and controlling the endless circumstances they faced. In fact, overall they were expected to have the wisdom of an attorney, counseling skills of a psychologist, writing abilities of an author, and physical skills of an athlete.

America's War on Crime

Lyndon Johnson established the Commission on Law Enforcement and the Administration of Justice in 1965. The Commission recommended in the 1967 publication, *The Challenge of Crime in a Free Society,* that police immediately make many operational changes and work toward numerous goals. These goals and standards called for monumental changes and implementation of many new programs to further professionalize law enforcement. Although some innovative and progressive police administrators initiated similar standards and goals at the state and local level, the impetus soon diminished. When fellow administrators did not join the movement, what had been accomplished soon faded.

The Law Enforcement Assistance Administration (LEAA) and the Omnibus Crime and Safe Streets Act were born from the Presidential Commission. America's "war on crime" was declared. Nearly $8 billion was directed toward the battle against crime from 1968 through 1978.[6]

When the Omnibus Crime and Safe Streets Act was passed in 1968, LEAA and the Federal Office of Law Enforcement Assistance merged. The Law Enforcement Educational Program (LEEP) was established to provide financial assistance for police officers to attend colleges and universities. All across the nation officers began to attend a variety of criminal justice and police science programs. The year 1969 marked the first year that LEEP funds were distributed; the relatively small sum of $6.5 million had been furnished. By the mid-1970s, the figure had risen to almost $40 million annually.

The social troubles of the previous decade had prompted extensive improvements within police departments. Education and training levels were advancing significantly. Fewer than 500 higher education programs participated in LEEP during its initial six months of operation, but the level of participation rose dramatically to 1,065 programs during 1975. Most college and university students participating in the program were employed by law enforcement agencies. More than 77,000 men and women attended college with LEEP funding during an average year. Grants were awarded to officers for payment of books, mandatory college fees, and university tuition.

During the latter part of the 1970s, LEEP funding gradually diminished. By the end of the program in 1981, the number of participating institutions had declined to just under 900. The reasoning behind such enormous federal funding was that a higher standard of education would lead to enhanced crime control. The government felt that improved working knowledge, expertise, initiative, and abilities would result in improved professionalism throughout the nation. The millions of dollars—though the subject of much debate—had a widespread, positive effect on the quality of college-level police education. Table 1.1 presents the nationwide education level in 1982.

TABLE 1.1
LAW ENFORCEMENT HIGHER EDUCATION, 1982

Education Level	*Percent*
Less than high school	00.7
High school	20.7
Less than associate degree	35.4
Associate degree	10.9
Junior year college or more	08.7
Bachelor's degree	13.4
Some graduate work	05.6
Master's degree	03.8
Law degree, doctorate, etc.	00.7

Source: Trinity Medical Center, Carrollton, Texas

TECHNOLOGICAL ADVANCEMENT

John Naisbitt's best-selling management text, *Megatrends*, refers to society's historical "waves." The first one was agricultural in nature. Its evolution occurred over thousands of years. The "second wave" dealt with America's Industrial Revolution. The basis of this era was natural resources, such as gas, oil, and coal. This time span has also passed. Naisbitt described the contemporary period as the "third wave," a time when society depends on technology and information systems to offset decreased natural resources. Law enforcement finds itself both the recipient of and dependent upon technology.

One of the first technological advancements in law enforcement came at the turn of the century. The Bertillon method of measurement was law enforcement's initial attempt to develop an accurate way to identify criminals. It was based upon the belief that literally taking detailed measurements of various physical characteristics, such as the dimensions of the ears and nose, could serve as a positive source of identification. Although archaic compared to contemporary fingerprint analysis, it was the best that could be done in a non-technology era. Since 1870 the Bertillon measurement method of identification was used to identify people arrested for crimes. In 1903 Will West was sentenced to Leavenworth Penitentiary. Upon his arrival at the penitentiary, a clerk discovered that another inmate with a similar name was already serving a life sentence there. The clerk was amazed when she realized that the two men were virtually identical in appearance. This incident destroyed the credibility of the Bertillon method and shifted tremendous significance toward fingerprint identification.

As America's technological advancements continued, so did those in law enforcement. The common use of transportation and communication resulted in corresponding improvements for the police. In the 1960s, law enforcement began to make great strides in adopting technology. Clerical functions and record systems were automated and computerized. Crime scene capabilities were enhanced. Forensic laboratories had relatively sophisticated equipment. Some agencies equipped patrol vehicles with computer-based information systems. Others had computer-assisted dispatching centers.

The National Crime Information Center (NCIC) was developed in 1965, and by 1967 it was operational across the country. This nationwide computerized information system provided officers with a means to communicate information concerning wanted persons and stolen vehicles or property. Similar state and local systems were gradually developed following the relatively successful implementation of NCIC.

In addition to educational funding, the Law Enforcement Assistance Administration provided millions of dollars to equip agencies across America with advanced technology. For example, in the 1960s patrol officers were armed with transmitters-receivers, allowing them to leave their vehicles without losing communications.

IN-SERVICE TRAINING

How well an officer is trained can make the difference between life and death. Yet, internal training has not been a strength of American law enforcement. Decades ago officers were issued their badges, guns, and uniforms, then simply told to go out and put someone in jail.

The public's demand for professionalism in officers has steadily increased. The days when size and strength were the primary prerequisite for employment are long gone. The potential for a devastating civil suit accompanies officers every day, and civil law and "the street" accept no excuses and allow no second chances. Standards of training have never been at such a demanding level.

Improper, outdated training—or the lack of it—is almost always a major allegation in civil suits against the police. Previously administrators did not perceive training of officers to be justifiable, when compared to other priorities and the difficulties associated with providing it. Now, however, they pay much closer attention to training needs within their agencies.

Were chiefs and sheriffs right when they felt there were many training-related problems? Is it more difficult for a police department to conduct training than a corporation? The answer is, Yes. Major challenges are associated with in-service police training, and it is more difficult for a law enforcement agency to train its officers than for a company to train its employees.

Effective, regular in-service training is necessary in all law enforcement agencies. The benefits have been reviewed. Now we will learn about the challenges trainers face.

First, officers assigned the responsibility for training often are provided with inadequate support. Most departments today have fewer than 25 full-time officers. Budgets do not always allow for a full-time detective, shift supervisor, or the community-oriented policing officer. It is certainly frustrating to be overworked and underpaid, yet not having the necessary knowledge or skill due to a lack of training can be devastating to an agency.

It is virtually impossible to train officers when you do not know how. In the past, few departments have understood how to develop or administer training programs, and many efforts have failed.

Second, insufficient budgets frequently have prevented the proper education of trainers and the provision of needed equipment. Each municipality has a limited revenue and tax base, and the administrators are responsible for establishing fiscal priorities. Unfortunately, employee training and development has been a low priority for many administrators.

Third, rotating shift schedules frequently used by departments create severe logistics problems. A variety of training schedules can be used to overcome this challenge. Some agencies overcome this obstacle well, primarily due to the support and cooperation of management.

Fourth, manpower shortages often interfere with training. Often there is too much work for officers to "break away" for training. Most agencies feel the frustration of having too few officers.

"No one has more power to impact the quality of life in a community more so than the uniformed patrol officer. Politicians can enact laws, but they have to rely on "street grunts" to enforce them, and to apply them fairly. Newspapers can praise or condemn, and focus public attention, but the resolution of the problem often lies in the hands of the beat cop. The cop doesn't have to wait for approval, or seek permission from a supervisor, in order to act. He can tow the abandoned cars, cite the reckless driver, and order the inconsiderate neighbor to turn down his music or face arrest. He can provide safe passage in the presence of the neighborhood bully, and remind everyone by his presence that we live in an ordered society, one that expects a certain standard of conduct. Done right, there is no more honorable way of making a living."

Timothy M. Dees
Law Enforcement Trainer and Author

THE ROLE OF THE POLICE

Most officers see themselves as crime fighters. Retiring officers' best memories may be about high-speed pursuits, shoot-outs, fights, or chasing criminals on foot. Rookies, similarly, may enjoy "putting the bad guy in jail." However, officers' roles have become complicated. Society expects them to handle virtually any situation. Further, they must manage complicated interpersonal relationships. These internal and external influences can affect their decisions. Pressure may come from relationships with friends and family; the local media; observing departmental regulations; local citizens; and personal ethical and moral beliefs. A role conflict may occur when:

1. Officers are confronted with expectations that are incompatible with their own beliefs.
2. Officers believe that others have different expectations about their actions.
3. Officers learn that others have different expectations for them.
4. Officers believe their role includes expectations, which may be contradictory or incompatible to the role established for them.

While each person has a somewhat different impression of the nature of the police function, based primarily upon personal experiences and contacts with police officers, there is a widespread popular conception of police reported by news and entertainment media. Police have come to be viewed as a body of people continually engaged in the exciting, dangerous, and competitive enterprise of apprehending and prosecuting criminals. This emphasis has led to a tendency on the part of both the public and the police to underestimate the range and complexity of the total police task. Police officers assigned to patrol duty in a large city are typically confronted with, at most, a few serious crimes in the course of a single tour of duty. Such involvement, particularly if there is some degree of danger, is viewed as real police work.

Officers, though, spend considerably more time keeping order, settling disputes, finding missing children, and managing drunks than responding to criminal conduct serious enough to call for arrest, prosecution, and correction. Officers, thus, perform a wide range of other functions—of a highly complex nature—often involving difficult social, behavioral, and political problems. [7]

Numerous studies before and after the President's Commission have reached a common conclusion: fighting crime is merely one of the roles of American police officers. The true role is a mixture of keeping the peace, serving the community, and fighting crime. Officers across the nation spend considerable more time providing community service than they do arresting criminals. Many officers prefer to think of themselves as protectors of their communities, but they are usually providers of community service.

Most citizens want their police force to do more than arrest criminals. They want to be able to call someone in the middle of the night to assist an elderly lady who has fallen out of bed. They want the police to help open their vehicles after locking the keys inside. People need the police to do these things because usually there is no other facet of society readily available and certainly none that are free. Many communities feel that their officers are always available for such tasks. However, while some departments restrict the amount of services for minor calls, the public still perceives law enforcers as public servants.

Developing a community service officer program or implementing a full community-oriented policing program is one alternative to the traditional peace keeping, crime-fighting role of police. It requires good budgetary management. In addition, it strengthens manpower because regular officers will have more time to concentrate on more serious matters. Community service officers require less training since they do not respond to high-risk or complicated situations. Establishing this type of program is fairly easy and allows agencies to provide the best possible service to the community.

THE CONTEMPORARY COP

Some officers today view themselves as crime fighters who have to handle menial tasks that should not be their responsibility. These officers are often frustrated by minor calls they believe are a waste of time. Although not all officers feel this way, if a substantial portion of an agency does, the department will provide a lower level of service.

Agencies must manage themselves to fit the role of providing a wide range of community services. Misdirected operations cause less efficiency and effectiveness. Obviously, the ultimate responsibility for directing a department lies with the chief administrator. Yet every officer has a responsibility to work toward the benefit of his or her community.

Continuing the "community service" attitude is easier said than done. Today's officers face more pressure than any of their predecessors. They are bombarded with increasing social, legal, and personal pressure. Many develop negative outlooks. Court decisions, community interaction, and politics—both internal and external—increase the relentless stress.

Stress affects all officers. They attempt to remain professional, yet the frustration of the judicial system and problems within agencies can be overwhelming. Further, they must face the worst of society. Sometimes dealing with human conflict, sadness, or despair can cause officers to feel powerless to uphold their responsibilities. The end result is sometimes excessive use of force, greed, or some form of unethical conduct.

Too much stress can harden officers' emotions. Out of necessity, officers may shield themselves from the misery that surrounds them. Unfortunately, by doing so they may sacrifice the compassion that is so essential to the performance of their duties. Suspicion and cynicism soon become psychological defenses for their actions. They also take "hardened" personalities home each day. These factors alone can make being a cop in contemporary America more dangerous mentally than physically.[8]

To an extent, officers' roles, amount of stress, and other aspects of their jobs vary from area to area. Differences exist from town to town and between zones in the same jurisdiction. Stress can come from many types of assignments, such as a long and boring stakeout, an uneventful midnight shift, or the pressures of working a high-crime district. "Survivors" learn how to take things in stride and not let the pressures of the street affect them.

No magic solution exists for America's crime problem. Recent developments with COMSTAT show great promise, but only time will tell. It was begun in 1994 by the New York City Police Department as a new aggressive form of working the street. This new approach is based upon an assertive philosophy of being proactive about crime prevention and constantly monitoring statistics. It then requires that patrol supervisors be held strictly accountable for preventing crime.

Technology will continue to advance, particularly with regard to computers. The new national NCIC 2000 program allows street officers to receive almost immediate responses to criminal history and other inquiries. Despite these advancements, officers themselves are law enforcement's greatest resource. They take pride in being innovative and progressive and setting the future of policing.

Great cops who become great leaders assume a tremendous amount of responsibility. They are responsible for preparing and leading employees by instructing, demonstrating, monitoring, and evaluating their progress. Their responsibilities extend not only to employees, but also to the organization, community, and profession as well.

Line supervisors are the ideal "vehicle" to instill ethics knowledge, skills, and abilities. Direct supervisors are employees' most influential role models. They shape and mold the beliefs and attitude of the work force—the long-lasting effect of what integrity-filled leaders say and do is astounding.

The Commandments of Leadership

If you do well, people will accuse you of ulterior motives.
Do well anyway.
If you are successful, you win false friends and true enemies.
Succeed anyway.
Honesty will make you vulnerable.
Be honest anyway.
The smallest men with the smallest minds can shoot down
the biggest men with the biggest ideas.
Think big anyway.
What you spend years building may be destroyed overnight.
Build anyway.
People really need help, but may attack you if you help them.
Help them anyway.
Give the world the best you've got,
knowing you may get kicked in the teeth.
Give the world the best you have anyway.

—Anonymous

LEADERSHIP

The history of police leadership has met with varied success. Chief administrators find that they have management responsibility to a municipal or county government, but are accountable for leadership of a paramilitary organization.

Leadership: The process through which people motivate, direct, influence, and communicate with those they work with to get them to perform in ways that will help the organization achieve its goals.

Power: The ability to influence others in an organization. Having many levels and sources, it is exercised in many ways and has led to the development of a variety of leadership techniques and theories.

For the ethical aspect of a great cop to become a way of life, it must be supported by strong leadership commitment and grow, upon a leadership style that promotes respect, fairness, and honesty. Power must be used to remove the obstacles that prevent others from doing their job.

Most good cops consider their formal advancement within their department and look ahead to becoming supervisors. The quicker you start developing your skills, the better prepared you will be when it is time to take the promotional examinations.

Obviously, line supervisors are essential to integrity becoming a way of life throughout an organization. They actually translate goals and objectives into results. They can make integrity a joke or a crucial requirement of every activity. The key is to deserve and earn their respect, trust, and support. Line supervisors must be capable of carrying out various responsibilities and using several skills.

Effective Human Relations Skills

It is important for all leaders to have the ability to apply respectful and fair human relation skills. Likewise, it is important to have the ability to assess the need for specific types of leadership by understanding an organization and the necessary skills.

In a leader, personality traits that perpetuate positive human relations are essential. The degree of effective communication in any leader is influenced greatly by the relationship with fellow workers.

Role Modeling

Supervisors act as trainers, counselors, and mentors for all employees. As a result of their constant contact with officers, they become major role models. Thus, it is vital for leaders to develop traits such as sincerity, loyalty, honesty, respect, and dedication. Their ability to influence and serve as role models is their greatest single source of power as leaders. For example, if line supervisors were unethical, it would be impossible for a company, association, or agency—let alone its officers—to be filled with integrity.

Sound Counseling Skills

Supervisors need the ability to counsel ethically in order to assist employees with a variety of professional and personal problems. These problems can be overcome through the ability to adjust to new circumstances, effective problem solving skills, and making sound ethical choices. Counseling strategies commonly used are emphasizing, suggesting, referring, reviewing, motivating, clarifying, informing, and interpreting. Leaders must convey integrity in the counseling process, regardless of situations or strategies.

Effective Motivation Skills

Supervisors must thoroughly understand the effect of motivation on employees in order to use motivation strategies that enhance successful performance. By far, the greatest motivator is helping others achieve worthwhile goals.

Leaders should emphasize enriching the work environment and positive, supportive relationships. In addition, they should remove obstacles that prevent officers from accomplishing their potential, whenever possible.

Exceptional Communication Skills

Supervisors need good reading, listening, writing, and speaking skills in order to ensure that trainees also have them. Supervisors need to be able to offer criticism constructively, disagree assertively, listen effectively, summarize messages correctly, and confirm final decisions accurately.

Communication is often a serious problem, whether within a marriage or a company of 700,000 people. Therefore, it must be given careful attention and stressed at all levels.

Effective Teaching Techniques

Supervisors should possess the instruction skills necessary to effectively teach officers. They do most of the training, yet few have ever been taught how to train. It would be ideal for supervisors to have a thorough understanding of the psychology of teaching, or have the ability to enhance concentration, comprehension, and learning retention. While this is not always possible, supervisors should at least attend a "train-the-trainer" course in an effort to provide effective training.

Accurate Evaluation Skills

All leaders must accept the challenges and responsibilities associated with fairly documenting behavior. If necessary, they must also effectively correct improper behavior. Evaluation should be carried out and communicated daily and recorded in writing honestly and respectfully.

Current Knowledge

Great leaders understand the never-ending need for knowledge. They also understand the need to remain current on strategies, information, and technology in their field. For police officers, the same is true, because leaders' lack of knowledge can discredit supervisors and entire organizations.

CONTEMPORARY LAW ENFORCEMENT AGENCIES

In the United States there are nearly 16,000 municipal, county, and state law enforcement agencies. This number includes approximately 12,500 police departments, 3,400 sheriff offices, and 49 state police agencies. The Bureau of Justice Statistics, a division of the Department of Justice, prepares an extensive profile of local law enforcement organizations in the United States every several years. The statistics presented here are gathered from the most recent report, as of January 2000. All agencies having more than 100 full-time, sworn officers are sent the detailed questionnaire every three to four years. The latest statistics, released in January 2000, were collected in June 1997.

The report is the only national research focusing on a wide assortment of vital information. It is a facet of the Law Enforcement Management Administrative Statistics program (LEMAS). The LEMAS survey, in addition to being interesting reading, yields valuable insight for every officer in the nation. The Bureau of Justice Statistics deserves great recognition for their tireless efforts in the development of such a challenging publication.

SUMMARY OF DATA ABOUT LAW ENFORCEMENT PERSONNEL

The law enforcement work force has changed during the last few years. Specifically, the number of officers throughout America has increased approximately 3% every year between 1993 and 1997. Between 1987 and 1993 the rate of growth had averaged only 1% each year. In addition, there has been a substantial increase in the number of women and minority officers since the mid-1990s. For municipal officers, ethnic and racial groups comprised 21.5% of all full-time officers in 1997, while they comprised only 19.1% in 1993. In sheriffs' departments the percentages were 19% in 1997 and 13.4% in 1993.

Table 1.2 presents various data on the police work force, salaries, operations, policies, and programs.

Table 1.2
Data About Law Enforcement Personnel

Work Force

Number of Agencies with 100+ Officers	*County Police*	*Municipal*	*Sheriff*
1,000 or more	7	31	14
500-999	7	30	15
250-499	12	72	38
100-249	7	278	69

Number of Employees and Sworn Officers	*County Police*	*Municipal*	*Sheriff*
Total number of employees	30,330	250.785	95,799
Total number of full-time, sworn officers	23,346	194,373	65,176

Average percent of	*County Police*	*Municipal*	*Sheriff*
Full-time employees who are sworn	79%	78%	71%
Officers assigned to respond to service calls	61%	63%	40%

Percentage of Sworn Employees by Job Function	*County Police*	*Municipal*	*Sheriff*
Administrative	6%	5%	5%
Field Operations	89%	90%	55%
Technical Support	4%	4%	4%
Jail Operations	1%	0%	27%
Court Operations	0%	0%	8%

Percentage of Civilian Employees by Job Function	*County Police*	*Municipal*	*Sheriff*
Administrative	14%	10%	11%
Field Operations	12%	13%	7%
Technical Support	68%	68%	38%
Jail Operations	1%	4%	38%

Percentage of Sworn Employees by Sex	*County Police*	*Municipal*	*Sheriff*
Male	90%	92%	86%
Female	10%	8%	14%

Percentage of Agencies With Educational Requirements for New Officers	*County Police*	*Municipal*	*Sheriff*
4-Year College Degree	0%	1%	1%
2-Year College Degree	9%	6%	2%
Non-Degree College Requirement	3%	14%	6%
High School Diploma	88%	81%	92%

continued

Percentage of Agencies with Residency Requirement for New Officers	*County Police*	*Municipal*	*Sheriff*
Within State	18%	5%	6%
Within City or County	12%	26%	42%
Within Other Specified Area	6%	19%	7%
No Requirement	64%	50%	45%

Median Number of Training Hours Required for New Officers	*County Police*	*Municipal*	*Sheriff*
Classroom Training Hours	704	640	480
Field Training Hours	400	480	400

Percentage of Agencies with Drug Testing of Civilian Employees	*County Police*	*Municipal*	*Sheriff*
Mandatory Testing of All Civilian Employees	0%	7%	10%
Random Selection Process	18%	9%	13%
When Use is Suspected	58%	44%	40%
No Drug Testing of Civilian Employees	39%	48%	45%

Percentage of Agencies with Drug Testing of Applicants for Sworn Positions	*County Police*	*Municipal*	*Sheriff*
Mandatory Testing of All Civilian Employees	70%	67%	66%
Random Selection Process	12%	2%	3%
When Use is Suspected	9%	7%	4%
No Drug Testing of Civilian Employees	18%	31%	32%

Salary

(As determined by the *Police Labor Monthly*, Salary Tracker Program, as of May 1999.)

*National Average Salary**	
Overall, Minimum for Police Officer	$33,272
Overall, Maximum for Police Officer	$42,896
Overall, Minimum for Detective	$45,497

Salary

(As determined by the LEMAS survey)

Average Base Starting Salary	*County Police*	*Municipal*	*Sheriff*
Chief Executive	$67,000	$62,600	$70,600
Entry-Level Officer	$26,000	$26,700	$23,400

Percentage Providing Special Pay for Officers	*County Police*	*Municipal*	*Sheriff*
Educational-Incentive Pay	33%	70%	62%
Hazardous-Duty Pay	42%	25%	34%
Merit Pay	45%	30%	42%
Shift-Differential Pay	58%	43%	28%

Percentage Allowing Membership In Unions	*County Police*	*Municipal*	*Sheriff*
Non-Police Union	6%	11%	12%
Police Union	64%	63%	44%
Police Association	48%	51%	36%

Percentage Supplying Cash Allowance for Body Armor	*County Police*	*Municipal*	*Sheriff*
Armor Supplied to All Regular Field Officers	85%	73%	76%
Armor Supplied Some Regular Field Officers	3%	5%	5%
Cash Allowance Given to All Regular Officers	6%	8%	3%

Operations

Percentage Using Specific Units	*County Police*	*Municipal*	*Sheriff*
Automobile Patrol Units	100%	100%	100%
Motorcycle Patrol Units	72%	61%	29%
Foot Patrol Units	31%	47%	5%
Bicycle Patrol Units	38%	40%	9%
Horse Patrol Units	16%	17%	8%
Boat Patrol Units	34%	12%	45%

Average Percentage of Patrol Units Comprised of	*County Police*	*Municipal*	*Sheriff*
One-Officer Patrol Units	97%	89%	95%
Two-Officer Patrol Units	3%	11%	5%

Percentage Requiring Body Armor to be Worn	*County Police*	*Municipal*	*Sheriff*
All Regular Officers	24%	29%	36%

continued

Percentage Authorizing the Use of Non-Lethal Weapons	*County Police*	*Municipal*	*Sheriff*
Baton, Collapsible	36%	48%	64%
Baton, PR-24	48%	58%	68%
Baton, Traditional	79%	64%	62%
Capture Net	0%	6%	1%
Carotid Hold	18%	20%	16%
Choke Hold	6%	4%	5%
Flash/Bang Grenade	61%	57%	63%
Pepper Spray	70%	69%	66%
Rubber Bullet	9%	9%	12%
Soft Projectile	9%	7%	10%
Stun Gun	6%	14%	28%
Tear Gas Personal Issue	42%	31%	35%
Tear Gas Large Volume	64%	42%	41%
Three-Pole Trip	3%	0%	0%

Policies and Programs

Percentage Agencies with	*County Police*	*Municipal*	*Sheriff*
Written Policies about Citizen Complaints	94%	99%	91%
Civilian Complaint Review Boards	12%	17%	17%

Requirement that Excessive-Force Complaints be	*County Police*	*Municipal*	*Sheriff*
Investigated by Outside Agency	58%	59%	60%

Percentage of Agencies Providing Right to Administrative Appeal in Excessive-Force Cases for	*County Police*	*Municipal*	*Sheriff*
Officers	100%	97%	91%
Citizens	42%	49%	42%

Percentage of Agencies in Which Final Disciplinary Decision on Excessive-Force Complaints Rests with:	*County Police*	*Municipal*	*Sheriff*
Chief, Sheriff, etc.	88%	82%	92%
Mayor, Commissioner, etc.	6%	9%	1%
Supervisory Personnel	0%	1%	4%
Other	6%	8%	3%

*From a regional perspective, wage increases exceeded 5% in the northeast and western north central area of the nation. The West, which has usually had the highest salaries, increased by slightly less than 4%. From the perspective of salaries related to the size of cities, the greatest increases were seen in communities with populations of 500,000 to 780,000 citizens, a 7.17% increase. Communities in the 84,000 to 89,000 population also averaged a 7% pay increase.

Information about subscriptions to *Police Labor Monthly* can be obtained by calling 409-291-7981.

Source: Brian Reaves, Ph.D., BJB Statistician, and Pheny Smith, Ph.D., BJB Statistician, Law Enforcement Management and Administrative Statistics, 1997: Data for Individual State and Local Agencies with 100 or More Officers, January 2000.

END NOTES

1. George G. Lillinger and Paul F. Cromwell, Jr., *Issues in Law Enforcement,* Holbrook Press, Boston: MA, 1975, p. 43.
2. Donald O. Schultz, *Special Problems in Law Enforcement,* Charles C. Thomas, Publisher, 1971, pp. 43–44.
3. Ibid., p. 45.
4. William Bopp and Donald Schultz, *Principles of American Law Enforcement and Criminal Justice,* Charles C. Thomas, Publisher, 1972, p. 35.
5. National Advisory Commission of Criminal Justice Standards and Goals, *Report on Police,* 1973, p. 380.
6. William Bopp and Donald Schultz, p. 35.
7. "The President's Commission on Law Enforcement and Administration of Justice, Task Force Report," The Police, U.S. Government Printing Office, Washington D.C., 1967, p. 13.
8. Paul Harvey, "The Most Dangerous Job: Law Enforcement," *The Los Angeles Times Syndicate,* 1986.

What it takes to be a good cop:

- **First, and most importantly, the officer must love and have a burning desire to help, assist and understand people.**
- **The officer must be fair and honest in all dealings with citizens regardless of their ethnic background or influential power within the community.**
- **The officer must be an example to all—morally, physically, spiritually—a first class role model.**
- **The officer must be mature and make wise decisions from knowledge learned.**
- **The officer must be willing to sacrifice self to assist and help people.**

What a cop is not:

Selfish
Conceited
Arrogant
Macho
Overbearing
Aloof
Uncaring
Undisciplined
Condescending
Lazy
Unapproachable
Haughty
A mind set of 'us vs them'
Hollywood enforcement

Larry Plott
Former Director Peace Officer Standards
and Training Commission, Idaho

CHAPTER 2

INTEGRITY VERSUS CORRUPTION

"The best and the worst people I have ever known were cops. There is something about police work that brings out extremes in people."

Timothy M. Dees
Law Enforcement Trainer and Author

Mediocrity is, in general terms, the opposite of professionalism. Great cops do not settle for mediocre careers or convince themselves that they are professional when they are not. They should rise above mediocrity and reassess their view of professionalism.

Professionalism is easy to recognize but extremely hard to define. Professionals pose particular personality traits and a professional attitude in an uncompromising pursuit of excellence.

Professionals, no matter what their endeavors, have a certain uniqueness in their attitudes. They take pride in the quality of their work, whether or not they are being observed or evaluated. Great cops, likewise, reflect this same trait.

Having this type of attitude has absolutely nothing to do with a particular occupation or endeavor. A professional attitude cannot be awarded, adjudicated or bestowed. Many people within various occupations and endeavors are demanding to be referred to as professionals. Often these same individuals are too busy climbing a self-serving ladder of success to be sincere in their efforts to improve their organization. Apathy and a general insincerity toward work are just not compatible with being professional.

Professionals display uncommon tenacity when others give up. Their positive view and untiring devotion pull them through adversity. The greatest individuals in America have always had this attitude, but it cannot be taught. It requires a positive outlook, confidence, self-esteem, and courage. In fact, when people act, think, and work like professionals, they truly must be professionals.

Ethics is a code or system of conduct and values with moral obligations and duties that define how to act. It is the training area needed most in many professions. Values are the basic beliefs that also guide actions and attitudes.

Most professions typically have done a poor job in preparing employees to make challenging ethical decisions on the job. This is incredible, especially in law enforcement, considering virtually every substantiated scandal and many civil suits have resulted from an unethical decision made by the parties involved.

CORRUPTION

To dedicated officers, reading a newspaper headline such as "police officer indicted during drug probe" is devastating. Every profession has members who violate moral, ethical, or professional standards of conduct. When the individual in question is a police officer, however, the offense seems even worse. Perhaps this is because no other occupation is afforded so much authority and responsibility. Citizens have practically given police officers the right to be judge, jury, and executioner. They expect high standards and offer little sympathy for ineptness or corruption.

Like the general public, "good" cops do not tolerate internal corruption. Law enforcement cannot prevent some officers from "going bad." Still, past and current levels of graft and corruption cannot be tolerated, and agencies must give no compassion to officers who have yielded to temptation. Substantial offenses should result in automatic termination following due process as standard policy across America.

It is essential to gain insight as to how departments can prevent problems of graft and corruption. We must learn from past mistakes and attempt to achieve unyielding, high standards for the future. The future is certain; what we make of it is not.

Study: Miami Police Department

One way to protect future standards is to examine how some departments have become infiltrated with corruption. As an example, more than seventy Miami city police officers were arrested between 1980 and the end of 1986. Chief Clarence Dickson of the Miami department wrote, "Paranoia and suspicion has run rampant through the police department and city hall, to the extent that free verbal expression cannot be exchanged without fear that the halls, telephones, desks, walls, and offices of everyone who is part of the decision-making process are illegally bugged."

"The inside of the Miami Police Department is filled with suspicion and uneasiness. Officers must live with the fact that they neither respect nor trust many fellow officers. Some officers have conducted major drug dealing. Others have been charged with murder. The Special Investigation Section has found $150,000 missing from its safe. Several hundred pounds of marijuana are also missing."[1]

The purpose here is not to examine isolated incidences of officers who have gone astray, but to inspect what went wrong with the organization. Our inquiry should be taken in the context of learning to safeguard against further similar tragedies. The extracts throughout this section are provided to alert you to the thought processes and consequences of corruption within law enforcement.

Demoralized, ashamed, sickened, scared, and frustrated are accurate words for how some Miami officers have felt. Who's to blame? Certainly the officers having committed unethical, immoral, or illegal acts. Yet what about supervisors who take part in or allow conversations that demean or ridicule administrators?

Aren't top-level managers who conveniently remain unaware of low morale or dissension within departments also to blame? Aren't they responsible for taking quick and decisive steps to correct departmentwide apathy? Could administrators be to blame for wide-sweeping internal policies that are blatantly unfair? Could local politicians be guilty of political interference or persuasion that demoralizes the rank and file?

The nightmare within the Miami force also involved racial and ethnic tension. White, African American, and Hispanic officers were openly angry and distrustful of one another. Separate bulletin boards for the various groups were displayed in the hallways. Some groups alleged that other groups hampered investigations of them. In addition, resentment over hiring practices and promotions tore fellow officers farther apart.

By responding to public pressures created by two devastating riots over a ten-year period, the city attempted to revolutionize the police department. Some officers considered that unprecedented recruitment and affirmative action efforts were beneficial and others harmful. Within two years, however, the department was transformed from a strong majority of white males to 60 percent minorities. It increased in size from 650 officers to 1,050.

The Miami agency had once been dominated by white males, then it suddenly found that white males comprised only one-third of the force. Now also 2 out of 5 Miami officers were Hispanic, women accounted for approximately 11 percent of the force, and almost 1 in 5 officers was African American.

If an agency has a majority of officers who are minorities, it does not mean the agency will be ineffective. The manner of the transformation is what went wrong. Such staggering changes did not occur without a price. Many veteran officers were convinced that they stood little chance for promotion. To them it appeared that efforts were being made to hire and promote only minorities. Thus, veteran officers at all levels were becoming discouraged by the changes.

In January of 1984, Chief Ken Harms was fired by Black city manager Howard Gary. Like most veteran officers, Chief Harms became frustrated when interdepartmental policies suddenly changed.

White officers who were irate over the number of minorities being promoted bombarded him. At the same time, the city manager demanded he promote more minorities. In a war in which there are no winners, Harms, a good cop, lost.

Herbert Breslow replaced Ken Harms as chief of police. Breslow was quick to follow the City Manager's recommendations to double his number of top administrators by including more African Americans, Hispanics, and a woman in the top echelon. Several civic leaders and politicians applauded the promotions, feeling that the department had finally come close to reaching the "recommended" integrated level of top administration.

However, there was ample reason to be angry over the promotions: seven officers were promoted from sergeant to major. In doing so, many lieutenants and captains were overlooked. Internal and political contacts appeared to be the overriding criteria necessary for these promotions. One of the promoted officers had been a leader in the African American benevolent association. Another was a former head of the Fraternal Order of Police. One female officer was the organizer of the women's officer group.[2]

The message was clear. If you were going to get anywhere within the department, it was who you knew that was going to get you there. Loyalty, dedication, and hard work were nice, but they didn't help you climb the ladder of success.

Affirmative action, being the right color or sex, playing politics well, or having friends in influential places had become the essential ingredients for success.

What could have been done to prevent the devastating political influence? Does the same thing happen in other governmental entities?

By January of 1985 Chief Breslow had been fired. Once again city politicians had forced the chief's termination. Clarence Dickson, an African American, replaced Breslow as the new chief of police. Though many problems occurred under Dickson's reign, the rank and file generally believed that the department's internal problems were not his fault. The city of Miami personnel division is another unit that repeatedly was the focus of criticism. In all fairness to city hall, the early 1980s was a period of incredible pressure and tension. First, there was the wave of Mariel immigrants (those who came to southern Florida in the early 1980s from the city of Mariel in western Cuba). Second, it was followed by the Liberty City riots. The riots had been ignited by the acquittal of five white metro officers who were accused of murdering an African American man. Because Miami continued to experience a very high violent crime rate, it was a logical conclusion that the city needed more police officers. Theoretically, the hiring of more minorities should have helped to improve racial tensions, but once again, the problem was the way they were hired and promoted.

Although personnel officials claim that hiring standards were never lowered, the force grew from 650 officers to 1,050 within two years. Every experienced chief of police or sheriff knows that hiring low-quality officers will result in low-quality performance. The consequences of superficial or indiscriminate hiring practices can be crippling.

During the early and mid-1980s, the Miami Police Department had a dark, unethical element within it. Fellowship and comradeship were replaced with animosity, resentfulness, and distrust.

Some citizens marked many dedicated officers with the label of "corrupt cop." Even so, there is no reason the Miami police can't rebuild. A "culture change" must replace distrust with respect and unity. Management must become totally committed to sincere, "people-oriented leadership."[3]

Study: New York City Police Department/Knapp Commission

The Knapp Commission, directed by Whitman Knapp as chair, conducted a monumental investigation into alleged corruption within the New York City Police Department. Mayor John Lindsey appointed the five-man commission in May of 1970, which completed a grueling, thorough investigation. Its findings can help others to understand why some officers yield to temptation; after all, every unethical act reflects on dedicated cops everywhere.

The Commission was responsible for three things:

1. To investigate input of corruption from the Commission's formation.
2. To evaluate New York City Police Department procedures concerning the investigation of corruption and to determine whether the procedures were adequate and followed with rapid and fair enforcement.
3. To recommend improvements for departmental procedures.[4]

The Commission soon was able to determine the extent of corruption and its findings were disturbing. Uniformed officers received regular payoffs from a variety of businesses and individuals. Detectives were routinely conducting shakedowns of individuals. Vice officers often received individual payoffs. Plainclothes officers frequently received semiweekly or monthly collections of payoffs from gambling organizations. Lastly, the assortment of bribes and payoffs was received not only from first-level officers, but sergeants and lieutenants as well. The percentage of corrupt officers still was relatively small, though the publicity surrounding the Commission did not emphasize it.

Following seemingly endless reviews of incidents, reports, transcriptions, and testimony, the Commission formed many conclusions. Many observers felt that underlying the Commission's logic was the premise that officers who become corrupt are simply "rotten apples" and little can be done to stop them. The Commission actually concluded the opposite. It concluded that a department trying to maintain a good public image frequently promotes the "rotten apple" theory. Instead, it stressed that managers should promote a realistic attitude toward corruption. Further, they should be honest, open, and factual in order to enhance the department's credibility and ability to deal with the causes of corruption.[5]

Contrary to the idea that corruption is difficult to prevent because a few "rotten apples" will be hired from time to time, the Commission determined that corruption can be curtailed by eliminating situations which expose officers to corruption. Open and honest internal communication also is crucial. Developing an atmosphere of trust, camaraderie, and loyalty is absolutely essential.

The Commission also made several more recommendations. Informal arrest quotas should be eliminated. Officers should always be reimbursed for legitimate expenses. Thorough hiring practices must be conducted. Personnel records should be centralized. Internal affairs divisions must also operate effectively. Lastly, relentless prosecution of officers who have succumbed to graft and corruption must occur, and internal investigation and outside assistance, when appropriate, should be used for investigation and prosecution. Above all, there should never be leniency.[6]

FACTS ABOUT BAD COPS

In 1996 without a lot of fanfare, the directors of Peace Officer Standards and Training Commissions and Councils throughout the nation were busy creating more work for themselves. With an untiring commitment, they researched data concerning all the officers within their state that had been formally disciplined by the Commission or Council between 1990 and 1995. When the statistics were compiled, they were forwarded to The National Institute of Ethics for nationwide analysis, conclusions, and recommendations to be developed.

Every state, in some way or another, responded—a 100 percent response rate. Such nationwide involvement has the potential to yield priceless new knowledge that can lead to preventing the devastation associated with brutality, corruption, and scandal. These facts could make training and leadership much more effective and efficient. A summary of the 85-page report follows on the next several pages.

It is the mission of the *National Law Enforcement Officer Disciplinary Research Project* to identify ways to prevent officer misconduct within law enforcement, based upon an extensive, accurate needs assessment.

Goals

I. Develop an effective survey instrument.
II. Obtain a response from 100% of the states.
III. Receive a written response of usable data from 50% of states.
IV. Ensure validity and reliability of survey findings.
V. Analyze the submitted data.
VI. Develop leadership and training conclusions and recommendations.
VII. Communicate findings throughout the nation.

Introduction

The importance of this research is profound for two reasons. First is the fact that ethics is our greatest training and leadership need. The second reason is that never before has there been a national law enforcement ethics-needs assessment that has focused on documented misconduct and the officers who committed the misconduct.

Findings

Number of Officers Disciplined by State Commissions/Councils

Findings: The total number of law enforcement officers having gone through the decertification process from 1990 through 1995 is 3,884. Of this number, 502 cases were dismissed, leaving 3,382 officers. Of the 3,382 officers, 2,296 officers were totally decertified. The term *decertification* means that the state in which an officer works has completed a process through which the state takes the officer's legal right to be a law enforcement officer away from the concerned officer. The person can no longer be an officer anywhere in that state.

When the number of cases dismissed (502) and cases still pending (278) are subtracted from the total number of 3,884 cases, we learn that there have been 3,104 cases where some form of discipline has been rendered.

Summary

Discipline Action	Number of Cases	Percentage of Cases
Revocation	2,296	59.1%
Case Dismissed	502	12.9%
Suspension	320	8.2%
Cases Still Pending	278	7.2%
Initial Certification Denied	244	6.3%
Probation	203	5.2%
Reprimand	41	1.1%

Conclusions:

1. The procedures which allow for the decertification of officers varies tremendously throughout the nation.
2. Decertification procedures should be standardized throughout the country.
3. State and nationwide systems for identifying officers who have been decertified should be developed and maintained.

Reasoning for Conclusions:

Some people should never be hired as law enforcement officers, because they have committed crimes or other unethical acts. Although the best solution is a hiring process effective enough that they are eliminated from consideration, the decertification process is the next best solution.

Many problems exist with contemporary decertification. Several states do not decertify officers. Many decertify for different reasons and in a variety of ways. Since there is no nationwide, standardized format used to track decertification data, many states had only data related to particular sections of the survey. This caused the survey sampling size to vary from topic to topic.

Age

Findings:

The average age of an officer who was the subject of this research was 32.

Conclusions:

The focus of the majority of contemporary law enforcement ethics training has been on academy, FTO training, or executive development. The fact that the average officer who has been decertified is 32 has reprioritized where the focus of ethics training should be—the officer with 5–10 years' experience.

Working within an organizational culture of disrespect and unfairness for several years can prompt officers to commit unethical acts.

Reasoning for Conclusions:

Many administrators and trainers presume that new officers are the most likely to give in to the temptations of anger, lust, greed, or peer pressure. As a result, ethics training has focused on new officers.

Sex

Findings:

The study revealed that of those officers who were the subject of this research, 93% were male and 7% were female.

Conclusions:

The percentage of decertified officers who are male and female is generally consistent with the overall percentages of male and female officers within the entire law enforcement profession.

Findings indicate that the female officers are slightly less likely to commit misconduct.

Reasoning for Conclusions:

According to the *Source Book of Criminal Justice Statistics,* 1995, as of October 31, 1994, males comprised 90.5% of sworn, full-time officers throughout the nation. Females accounted for 9.5%.

Race/Ethnicity

Findings:

The study revealed that of those officers who were the subject of this research:
73% were Caucasian
19% were African American
8% were Hispanic

Conclusions:

The percentage of Caucasian officers who have been processed for decertification throughout the nation is 8% less than the overall percentage of Caucasian officers within the work force.

The percentage of African American officers who have been processed for decertification is 8% higher than the overall percentage of African American officers within the work force.

The percentage of Hispanic officers who commit unethical acts is slightly less than the overall percentage of Hispanic officers within the work force.

Reasoning for Conclusions:

According to the *U.S. Department of Justice, Bureau of Justice Statistics,* 1996, the ". . . race and ethnicity of full-time officers in local police departments is White/80.9%, Black/11.3% and Hispanic/6.2%.

Education Level

Findings:

The study revealed that of those officers who were the subject of this research:

70% had a high school degree
11% had a GED
10% had an A.A./A.S. degree
9% had a B.A./B.S. degree

Conclusions:

The study concluded that there were no substantial differences of officers who were the subject of this research compared to the national education levels of the overall population of law enforcement officers. This could not be verified by confirmed statistics, however.

Number of Employments

Findings:

The study revealed that officers who were the subject of this research averaged 2.16 previous employments.

Conclusions:

The number of previous employments of officers who have been processed for decertification is not significantly different than the normal career changes of officers throughout the work force. This could not be verified by confirmed statistics, however.

Employment Status

Findings:

The study determined that of those officers who were the subject of this research, 92% are full-time, sworn officers, 5% are part-time, and 3% are auxiliary officers.

Type of Officer

Findings:

The study determined that the type of officers who were the subject of this research were comprised as follows:
56% were city officers, although they comprise 66% of the work force
33% were county deputies, although they comprise 25% of the work force
11% were state officers, although they comprise 8% of the work force

Conclusions:

The percentage of municipal officers who have been processed for decertification throughout the nation is 10% less than the overall percentage of municipal officers within the work force.

The percentage of sheriff deputies who have been processed for decertification throughout the nation is 8% greater than the overall percentage of municipal officers within the work force.

The percentage of state officers who have been processed for decertification throughout the nation is 3% greater than the overall percentage of state officers within the work force.

Reasoning for Conclusions:

According to the *Bureau of Justice Statistics, Local Police Departments, NCJ-148822,* Washington DC, 1996, there are 622,913 city, county, and state law enforcement officers in America.
Of this number, approximately:
66%, or 415,224, are city police officers
25%, or 155,815, are sheriff deputies
8%, or 51,874, are state police officers

Rank

Findings:

The study determined that 85% of officers who have been processed for decertification between 1990 and 1995 were patrol officers, deputies, or troopers.

Conclusions:

From the perspective of rank or position within a law enforcement agency, the rank of patrol officer, deputy, or trooper accounts for an unproportionally high number of officers who commit unethical acts.

In-service ethics training should target patrol officers as a high-priority focus.

Reasoning for Conclusions:

From the perspective of rank, the percentages of officers ultimately processed for decertification between 1990 and 1995 resulting from misconduct is presented below.

Patrolman	59%
Deputy	22%
Sergeant	5%
Trooper	4%
Detective	3%
Special Agent	3%

Captain, Chief, Sheriff, Lieutenant 1% for each group
Assistant Chief, Major, Wildlife Officer each less than 1%

Years Sworn

Findings:

The study revealed that officers who were processed for decertification from 1990 through 1995 had an average of 7.2 years of sworn service when the decertification was initiated.

Conclusions:

Officers most likely to commit unethical acts are not rookies, but those with 5 to 10 years of service.

Reasoning for Conclusions:

The fact that officers processed for decertification had an average of 7.2 years of sworn service is consistent with the fact that the average age of these officers is 32. This is a single, yet vital, fact from which ethics training can become more effective.

Offenses Charged

Findings:

The four most frequent crimes committed by officers who have been processed for decertification are *making false statements/reports* (19.92%), *larceny* (12.12%), *sex offenses other than rape* (9.48%), *battery* (9.15%). These four offenses comprise 51% of the crimes for which officers have been decertified. Other than filing false statements/reports, virtually all other offenses committed by the concerned officers can be grouped into four groups:

Greed (25.47%)
Larceny (12.12%)
Fraud/Forgery (5.01%)
Sale of Cocaine (3.08%)
Sale of Cannabis (1.36%)
Robbery (1.19%)
Bribery (1.19%)
Stolen Property (1.11%)
Gambling (.41%)

Anger (19.69%)
Battery (9.15%)
Excessive Use of Force (5.05%)
Weapon Offense (4.02%)
Family Offense (1.47%)

Lust (12.74%)
Sexual Offenses Other than Rape (9.48%)
Sexual Battery/Rape (2.77%)
Morals/Decency Crimes (.49%)

Peer Pressure (12.76%)
Driving Under the Influence (5.08%)
Drugs Other than Cocaine and Cannabis (4.64%)
Cocaine Drug Test (1.68%)
Cannabis Drug Test (1.36%)

Top Ten Offenses for Which Officers are Decertified:

	Offense	Percent
1.	False Statements/Reports	19.92%
2.	Larceny	12.12%
3.	Sex Offenses Other than Rape	9.48%
4.	Battery	9.15%
5.	Driving Under the Influence	5.08%
6.	Excessive Use of Force	5.05%
7.	Fraud/Forgery	5.03%
8.	Drugs Other than Cannabis/Cocaine	4.64%
9.	Weapon Offenses	4.02%
10.	Cocaine—Possession or Sale	3.08%

Conclusions:

Research should begin immediately to determine the root causes for the excessive number of false statements or reports that are committed.

Individual agencies should conduct their own ethics training needs assessments, focusing on the motivations of anger, lust, greed, and peer pressure.

Individual agencies should conduct their own leadership needs assessments, focusing on the motivations of anger, lust, greed, and peer pressure.

Academies, FTO programs, in-service training, leadership training, civilian training, and job-specific training should focus on anger, lust, greed, peer pressure, and falsifying records.

Reasoning for Conclusions:

Understanding training and leadership needs, both for professions or specific organizations, is crucial for effective training and leadership. It is very unlikely that training and leadership can be as effective if the needs of the agency are not known. Conducting a training needs assessment should be the first step in the development of any training program. The value of this research is that it is a nationwide needs assessment for law enforcement ethics training and leadership.

Recommendations to Stop Bad Cops

1. Standardize decertification terminology and procedures throughout the country. The lack of a single standardized process makes effective nationwide tracking of officers who have been totally decertified very difficult.
2. Develop and maintain state and nationwide systems for identifying officers who have been decertified as a means for preventing decertified officers from being hired unknowingly by agencies.
3. Identify a group of IADLEST* members who are willing to assist states seeking to develop or enhance decertification.
4. Continue to track and analyze decertification statistics so that we can continue to become more effective at preventing misconduct.

Recommendations for Leadership

1. Orchestrate a positive organizational culture within the patrol division—an extremely high priority.
2. Hold all levels of leadership accountable for being role models for integrity.
3. National and state sheriff's associations, state POST commissions, and law enforcement academies should assist state and county law enforcement agencies in providing state-of-the-art executive development ethics training.

---

* International Association of Directors of Law Enforcement Standards and Training

4. State police administrators and particularly sheriffs throughout the country should implement contemporary leadership solutions to prevent misconduct.

The state of the art for preventing officer misconduct is comprised of three steps:

- Maintain a leadership style driven by respect and dignity for all employees.
- Implement ethical dilemma simulation training to anchor an ethical decision-making process into the long-term memory of officers.
- Implement a comprehensive administrative process that prevents misconduct.

5. Administrators must embrace and support their Field Training Officer programs.

   Field training officers create the culture of patrol divisions and individual patrol shifts. If FTOs are cynical, bitter, resentful, or unethical, they will consistently develop new officers with the same outlook. On the other hand, if FTOs are the end result of effective FTO selection, training, compensation, recognition, and leadership, they will likely create a patrol culture of positive, motivated officers.
6. Administrators should conduct an ethics needs assessment to determine the integrity-related needs of their organization.

   The first step for preventing unethical acts is to determine your ethics-related needs. Identifying your needs should then be the basis for immediate and future training and leadership initiatives.

Recommendations for Training

1. Develop effective in-service ethics training that is focused on the 5 to 10-year patrol officer.

   The majority of contemporary law enforcement ethics training focus has been on academy, FTO training, or executive development. The fact that the average officer who has been decertified is 32 years of age has reprioritized where the focus of ethics training should be: the officer with 5–10 years' experience.
2. Address the ethical perspectives of the topic.
3. Develop ethical dilemma simulation training about the most common offenses committed by officers.[7]

HONESTY

Everyone agrees that honesty is a worthy principle. Every major religion preaches it, schools teach it, civil litigation enforces it, and businesses, governments, and individuals claim that they practice it.

Yet honesty is frequently ignored. Self-serving interests often rule out the choice to do the right thing. Few endeavors offer more temptations to be dishonest than law enforcement. Some officers finding an "open door" to a business at 0315 hours may be tempted to steal. Others may find it difficult to answer questions honestly during a deposition if the answer would make them look bad.

In a profession representing integrity, pride, protection, and service to others, why is there a problem with officers being dishonest? Unfortunately, it has become part of our culture in major and minor ways. For some, the "American way" includes cheating on your taxes, lying to avoid awkward situations, and as a child witnessing seemingly harmless examples of dishonesty set by role models. All professions must embed honesty in their operations and work force.

"Cops are expected to be loyal, but their loyalty is often misplaced. At the outset of their careers, they swear loyalty to the Constitution, to the laws of their state, and to the communities they serve. Later on, they are told that other loyalties are expected—to other cops, to certain supervisors and administrators, to some unwritten code of conduct. These are not necessarily bad things, until they conflict with that first loyalty—the one that they formally swore to, the one that is binding. There's a reason that one came first. Putting it second, or even farther down the list, is a betrayal of their office, and the public trust."

Timothy M. Dees
Law Enforcement Trainer and Author

ETHICAL PERCEPTIONS

Historically, regardless of the profession, some organizations have reacted to individual acts of misconduct or devastating scandals and have done little other than to hold employees accountable for their actions following a violation.

Others have knowingly ignored the constant lack of integrity by some employees, fearing that addressing the problem might ultimately result in negative publicity. Thus, they knowingly let it continue.

In the worst of cases, supervisors become driven by the hope that employees and the community will not learn about internal misdeeds. Meanwhile, an internal climate of dishonesty is perpetuated by the role modeling of deceitfulness and bitter internal politics. Ignoring it allows it to continue.

Good administrators and law enforcement leaders are inherently driven by their own integrity and honor. It is their personal role modeling of honesty and respect toward others that also molds a culture of dignity, integrity, and honor in those they manage or lead. Their leadership style prevents misconduct and maintains integrity within an organization.

The 1990s witnessed a revolution of knowledge and ability to prevent unethical acts. Most of the elements that comprise the current state of the art focus on prevention, rather than reaction after careers have been destroyed and reputations have been lost. Ethics has been law enforcement's greatest training need since the mid-1980s. Prior to this time, law enforcement's most significant need was firearms training. For decades instructors had been anchoring the wrong survival responses into officers' long-term memory during firearms training. Because research had not been conducted prior to the mid-1970s, firearms instructors did not know that they were sometimes anchoring detrimental behavior into the long-term memory of officers. As an example, after firing several rounds at targets during firearms training, officers were usually required to throw the empty shell casings into a bucket behind them. As a result, this response was anchored deeply into their long-term memory. This means that in a real shooting these same officers may take the time to throw the empty casing into a nonexistent bucket. This response caused some officers to be killed or injured.

The only reason this unfortunate situation changed was that the FBI in Quantico, Virginia, conducted a training needs assessment and literally determined how officers were dying. The results of this study changed firearms training forever. Armed with this vital new knowledge, trainers and corporations developed discretionary dilemma video training systems. The result has been dramatically improved firearms training. Similarly, identical improvements have been made with ethics training. Fortunately so, because virtually none of the 16,000 agencies across the nation had provided any in-service ethics training prior to the 1980s.

With the exception of dying in the line of duty, nothing is more devastating to an officer's personal life, leader's career, or an agency's respect than allegations of unethical conduct. Such allegations are the basis for many civil suits filed against law enforcement.

Many officers commit suicide as a result. Each year two to three times the number of officers who die in the line of duty commit suicide. Some do so because they made a foolish 3- to 5-second decision that would potentially ruin their lives, and they feel as though they have lost their career, dignity, respect, and retirement.

Such split-second, poor decisions—when facing moments of anger, lust, greed, or peer pressure—can destroy the future of good people. Thus, it is crucial that we help each other maintain a strong, positive, and ethical mental outlook.

EXCUSES

Some people believe that abiding by a strict, professional code of ethics is an unrealistic goal. Others feel that although high ideals and integrity are worthy objectives to seek, they remain impractical in real life.

They claim that officers will never be able to follow exceedingly high ethical standards because their values were already programmed by an unethical society. This thought is sometimes used as an excuse by corrupt officers who say it is all right to steal, lie, or cheat because the rest of society does it.

People who believe high ideals and ethical standards cannot be met by America's police officers are mistaken. They have not felt the brotherhood and camaraderie of dedicated street cops. Sincere cops are bound together in their moral convictions by the sweat and blood of years on the street. Their loyalties are to ethical principles supported by pride and guts. Their language is theirs alone. Unlike the few weak individuals among them, they neither search for nor need excuses. Their actions are above the temptations of money, lust, or drugs.

Officers who use self-serving excuses or do things morally or ethically wrong do it because their character is weak. There is little difference in why a child lies and why officers try to justify unethical acts by using deceit to explain their actions. A weakness of character caused them not to accept responsibility for their own actions.

Professionalism is tarnished every time an officer is intentionally misleading. Excuses are easy to think of. The officer who "fixes" a ticket merely says he was mistaken. Those who accept gratuities and favors from local businesses usually claim there is no harm in taking gifts.

After all, businessmen are simply trying to support their local police department. In reality, such businessmen frequently expect to receive special treatment in return.

In 1984, Miami police officer Carlos Pedrera was accused of committing criminal acts. During his corruption trial in 1987, Pedrera testified that he went from committing small-time rip-offs of cocaine smugglers to making more than $1 million in two boatyard drug deals. At the time of his testimony, Pedrera was facing twenty years in prison. He testified against three former Miami officers accused of a cocaine rip-off murder charge. When the U.S. attorney asked Pedrera why he did these things, Pedrera answered simply, "I needed the money."

Excuses like "I needed the money" or the acceptance of gifts and favors from local businesses as "signs of their appreciation" are nothing more than a cop-out. Vice detectives or other officers who have sex with hookers try to justify it by telling themselves they did not hurt anyone. Investigators who commit perjury may be trying to make themselves look good or convict someone they dislike. They might feel that "the scumbag deserves it."

Nothing is complicated about officers who cheat, steal, or lie. All of them realize that what they are doing is wrong, but their character is not strong enough to avoid it.

While ethics education and training will help a great deal, another solution is to not hire "weak" individuals to begin with. Officers who give in to daily temptations should not be cops. Therefore, hiring practices must be strengthened so that agencies can detect character weaknesses.

CODE OF ETHICS

Principles provide guidance, direction, and vision. For law enforcement, an example of such principles is the Law Enforcement Code of Ethics. A sense of professional responsibility can come from understanding the reasons the Code of Ethics was established.

Most people respect the regulations set to assist their professional life. A clear understanding and appreciation of the responsibilities needed to achieve professionalism can lay the foundation for a strong tomorrow. The first step in developing an appreciation is to know the Code of Ethics, and sincere reflection can help to understand its value.

In 1956 a committee of the Peace Officers Research Association of California drafted a Code of Ethics. After it was edited by a committee of the California Peace Officers Association, it was adopted that same year. It gradually became accepted throughout the nation as the Law Enforcement Code of Ethics. It was adopted by the International Association of Chiefs of Police in 1957. The code is shown in Figure 2.1.

FIGURE 2.1

Law Enforcement Code of Ethics

As a law enforcement officer, my fundamental duty is to serve mankind; to safeguard lives and property; to protect the innocent against deception, the weak against oppression or intimidation, and the peaceful against violence or disorder; and to respect the constitutional rights of all men to liberty, equality and justice.

I will keep my private life unsullied as an example to all; maintain courageous calm in the face of danger, scorn or ridicule; develop self-restraint; and be constantly mindful of the welfare of others. Honest in thought and deed in both my personal and official life, I will be exemplary in obeying the laws of the land and the regulations of my department. Whatever I see of a confidential nature of that which is confided to me in my official capacity will be kept ever secret unless revelation is necessary in the performance of my duty.

I will never act officiously or permit personal feelings, prejudices, animosities or friendships to influence my decisions. With no compromise for crime in the relentless prosecution of criminals, we will enforce the law courteously and appropriately without fear or favor, malice, or ill will, never employing unnecessary force or violence and never accepting gratuities.I recognize the badge of an officer as a symbol of public faith, and I accept it, as a public trust to be held so long as I am true to the ethics of police service. I will constantly strive to achieve these objectives, ideals, and dedication of myself before God to my chosen profession . . . law enforcement.[8]

OATH OF HONOR—THE NEW MOVEMENT

As the effort to incorporate ethics training fully within law enforcement begins, it will be important to heighten the visibility and awareness of ethics across the profession. A public affirmation to adhering to the current code of ethics and the adoption of an Oath of Honor will have to be undertaken along with role modeling and mentoring—powerful vehicles for changing behavior.

To be successful at enhancing integrity within an organization, leaders must ensure that ethical mentoring and role modeling is consistent, frequent, and visible. Therefore, the committee from the California Peace Officers Association wholeheartedly supported the creation of a symbolic reverberance and public affirmation in order to attest a commitment to ethical conduct. After numerous drafts and conferences, the following Law Enforcement Oath of Honor was recommended in 1997 as the International Association of Chiefs of Police (IACP) symbolic statement of commitment to ethical behavior. The oath reads:

On my honor,
I will never betray my badge,
my integrity, my character, or the public trust.
I will always have
the courage to hold myself
and others accountable for our actions.
I will always uphold the constitution and community I serve.

An oath is a solemn pledge one makes and intends to follow. Before officers take the Law Enforcement Oath of Honor, it is important that they understand what it means.

Honor is one's word given as a guarantee.
Betray is the breaking of faith with the public trust.
Badge is the symbol of office.
Integrity is adherence to a code of honesty, in both private and public life.
Character is the set of qualities that distinguish an individual.
Public trust is a charge of duty imposed in faith toward those being served.
Courage is the strength to withstand unethical pressure, fear, or danger.
Accountable is being answerable and responsible to the oath.
Community is the jurisdiction and citizens served.

Because the oath is brief it can be constantly referred to and reinforced during conversations with FTOs and line supervisors and used in the following situations:

- Referred to by administrators while communicating with others.
- Placed on the back of academy students' name cards, so that they can refer to it as needed.

- Placed visibly in all police academies and law enforcement agencies.
- Enlarged and framed so that all academy students can see it.
- Recited at all official police ceremonies and gatherings.
- Printed on equipment labels.
- Used as a backdrop in citizen's meetings and news media events.

The oath reconfirms the significance of integrity within agencies. It helps bring the entire profession together to show that the vast majority of law enforcement officers are good, decent individuals willing to step forward to stop unethical acts by any members of the profession.[9]

CANON OF POLICE ETHICS

Great cops require deep and abiding sincerity, in addition to sensible action. Therefore, another priceless document in the law enforcement profession is the Canon of Police Ethics, in Figure 2.2, which provides ethical direction and guidance. Topics include individual conduct, limitations of authority, primary responsibility, duty to serve, and attitude.

FIGURE 2.2

Canon of Police Ethics

Article 1. Primary Responsibility of Job
The primary responsibility of the police service and of the individual officer, is the protection of the people of the United States through the upholding of their laws; chief among these is the Constitution of the United States and its amendments. The law enforcement officer always represents the whole of the community and its legally expressed will and is never the arm of any political party or clique.
Article 2. Limitations of Authority
The first duty of a law enforcement officer, as upholder of the law, is to know its bounds upon him in enforcing it. Because he represents the legal will of the community, be it local, state or federal, he must be aware of the limitations and proscriptions which the people, through the law, have placed upon him. He must recognize the genius of the American system of government, which gives to no man, groups of men, or institution, absolute power, and he must insure that he, as a prime defender of that system, does not pervert its character.
Article 3. Duty to be Familiar with the Law and with Responsibilities of Self and Other Public Officials
The law enforcement officer shall assiduously apply himself to the study of the principles of the laws which he is sworn to uphold. He will make certain of his responsibilities in the

particulars of their enforcement, seeking aid from his superiors in matters of technicality or principle when these are not clear to him; he will make special effort to fully understand his relationship to other public officials, including other law enforcement agencies, particularly on matters of jurisdiction, both geographically and substantively.

Article 4. Utilization of Proper Means to Gain Proper Ends

The law enforcement officer shall be mindful of his responsibility to pay strict heed to the selection of means in discharging the duties of his office. Violations of law or disregard for public safety and property on the part of an officer are intrinsically wrong; they are self-defeating in that they instill in the public mind a like disposition. The employment of illegal means, no matter how worthy the end, is certain to encourage disrespect for the law and its officers. If the law is to be honored, those who enforce it must first honor it.

Article 5. Cooperation with Public Officials in the Discharge of Their Authorized Duties

The law enforcement officer shall cooperate fully with other public officials in the discharge of authorized duties, regardless of party affiliation or personal prejudice. He shall be meticulous, however, in assuring himself of the propriety, under the law, of such actions and shall guard against the use of his officer or person, whether knowingly or unknowingly, in any improper or illegal action. In any situation open to question, he shall seek authority from his superior officer, giving him a full report of the proposed service or action.

Article 6. Private Conduct

The law enforcement officer shall be mindful of his special identification by the public as an upholder of the law. Laxity of conduct or manner in private life, expressing either disrespect for the law or seeking to gain special privilege, cannot but reflect upon the police officer and the police service. The community and the service require that the law enforcement officer lead the life of a decent and honorable man. Following the career of a policeman gives no man special prerequisites. It does give the satisfaction and pride of following and furthering an unbroken tradition of safeguarding the American republic. The officer who reflects upon this tradition will not degrade it.

Article 7. Conduct Toward the Public

The law enforcement officer, mindful of his responsibility to the whole community, shall deal with individuals of the community in a manner calculated to instill respect for its laws and its police service. The law enforcement officer shall conduct his official life in a manner such as will inspire confidence and trust. Thus, he will be neither overbearing nor

continued

subservient, as no individual citizen has an obligation to stand in awe of him nor a right to command him. The officer will give service where he can, and require compliance with the law. He will do neither from personal preference or prejudice but rather as a duly appointed officer of the law discharging his sworn obligation.

Article 8. Conduct in Arresting and Dealing with Law Violators

The law enforcement officer shall use his powers of arrest strictly in accordance with the law and with due regard to the rights of the citizen concerned. His office gives him no right to prosecute the violator or to mete out punishment for the offense. He shall, at all times, have a clear appreciation of his responsibilities and limitations regarding detention of the violator; he shall conduct himself in such a manner as will minimize the possibility of having to use force. To this end he shall cultivate a dedication to the service of the people and the equitable upholding of their laws whether in the handling of law violators or in dealing with the law-abiding.

Article 9. Gifts and Favors

The law enforcement officer, representing government, bears the heavy responsibility of maintaining, in his own conduct, the honor and integrity of all government institutions. He shall, therefore, guard against placing himself in a position in which any person can expect special consideration or in which the public can reasonably assume that special consideration is being given. Thus, he should be firm in refusing gifts, favors, or gratuities, large or small, which can, in the public mind, be interpreted as capable of influencing his judgment in the discharge of his duties.

Article 10. Presentation of Evidence

The law enforcement officer shall be concerned equally in the prosecution of the wrongdoer and the defense of their innocence. He shall ascertain what constitutes evidence and shall present such evidence impartially and without malice. In so doing, he will ignore social, political, and all other distinctions among the person involved, strengthening the tradition of the reliability and integrity of an officer's word.

The law enforcement officer shall take special pains to increase his perception and skill of observation, mindful that in many situations his is the sole impartial testimony to the facts of a case.

Article 11. Attitude Toward Profession

The law enforcement officer shall regard the discharge of his duties as a public trust and recognize his responsibility as a public servant. By diligent study and sincere attention to self-improvement he shall strive to make the best possible application of science to the solution of crime and, in the field of human relationships, strive for effective leadership and public influence in matters affecting public safety. He shall appreciate the importance and responsibility of his office, and hold police work to be an honorable profession rendering valuable service to his community and his country.

Source: International Association of Chiefs of Police, "The Patrol Operation," Washington, DC, 1970, p. 41–50.

RESPONSIBILITIES

Law enforcement cannot be a profession unless its members have a strong sense of obligation to their responsibilities. Officers must accept the fact that their responsibilities extend far beyond the specific orders given by supervisors. All actions support dedication that transcends specific assignments to ethical principles. Only deep abiding commitment to integrity carries people through the rough times that are inevitable in a police career.

People all have basic desires and they all have responsibilities as well. They should be aware of their own strengths and weaknesses and know that it is possible to remain honest and succeed in a world that is often riddled with dishonesty. They need to have a healthy self-esteem and believe in themselves. Although stealing and lying will "reward" for a moment, they create "losers" and short-lived success.

Community Leaders and Role Models

Corporations must comply with high standards of self-imposed responsibility, standards based upon the understanding that the trust customers have in them cannot be betrayed. This responsibility includes integrity leadership.

The standard for policing must be even higher than that for corporations. Law enforcement is generally recognized as vital to every community. Officers, of course, have the same duties to their community as all citizens: to pay taxes, vote, participate in community activities, serve on juries, and live as civic-minded members of the local area. To fulfill their duties, police comply with a high standard of self-imposed responsibility.

Officers must accept their responsibility to provide leadership within their communities. They are perceived to be natural role models. When someone has a problem, needs advice, or seeks assistance, they are often sought, even when off duty. This is very similar to the way a clergy member is sought for advice and help—it is a role that cannot be ignored or taken lightly.

Member of Fellowship

True professionals realize they owe a sincere, dedicated, and loyal commitment to their profession. Regardless of title, assignment, or seniority, every sincere person is a member of an everlasting fellowship of integrity.

To promote professionalism is never corny or outdated. In fact, few feelings are more satisfying than giving your best effort for a worthy cause.

Relationship with Citizens

When veteran officers think back to when they were rookies they'll probably recall the main reason they wanted to be an officer was to help others. Law enforcement has been and always will be the profession that people call upon most often when they really need help. Even people with low morals—who would not hesitate to curse at officers—scream and shout for their help when in serious trouble. All officers have an obligation to give every citizen the same level of professional help. These moments can determine if officers have what it takes to be true professionals.

It's easy to understand why some officers become hardened and cynical after several years of "working the street;" they can lose sight of the fact that most people are still decent individuals. However, they often meet good people who are going through one of their worst moments: after an emotional car accident, at their home during a rare family fight; or as a victim of theft, burglary, or assault, who is not in a particularly good mood. Times of tragedy are sometimes the only reason people meet an officer or deputy.

Most professions have established terms that explain the relationship between the professional and their "client." Relationships of physician/patient, lawyer/client, and teacher/pupil are logical and easily understood. In these situations confidence and trust exist, and a bond of respect is directed toward the people providing the assistance and guidance.

Similarly, trust, sincerity, and confidence can be built between citizens and law enforcement agencies. This is the basis for community-oriented policing. Every police department has a responsibility to the citizens it serves, without exceptions. Honesty, integrity and respect should be an oath of integrity leadership for all leaders.

Most citizens form an opinion of their local law enforcement based on one or two isolated incidents with individual officers. Frequently individuals' respect or lack of respect stems from incidents that occurred years ago in other communities.

TEMPTATIONS

At one time or another, all officers face seemingly endless situations during their careers. Although there is no place for weak character—and few are so weak that they submit to major thefts, drug dealings, or bribery—one bad officer is one too many.

Common, everyday situations afford choices between right and wrong and offer temptations that are tempting enough that some officers try to justify doing something that they know is wrong. The following information provides insight and guidance for surviving such difficult situations.

Excessive Force

We have all seen newspaper headlines or television segments about police brutality. Many experienced officers recall situations in which a fellow officer used excessive force when it was unnecessary. It is difficult to determine what is "excessive force;" what may be considered excessive by some officers is justifiable in the minds of others.

We all know the difference between right and wrong. It is wise to follow the meaning of the Golden Rule: Do unto others as you would have them do unto you. Then ask yourself what you would do if your family were standing beside you.

Officers who have trouble controlling their anger or show a tendency toward revenge should seek counseling if they cannot resolve it themselves. If the problem continues, the best solution would be for them to find employment in another endeavor.

As an officer, what would you be seeing if you saw another officer take a few extra hits to an individual with a nightstick, PR-24, or asp? You would be watching him or her lose a career, right in front of you. If this ever happens, stop the officer. This probably will not be easy, as the officer may be asking you to join in the beating. However, it is one of those moments when your true character will prevail: have the guts to stop the excessive force. Be very careful, though, to never lose control of the citizen, who may still be a danger to all the officers present.

Gratuities

A gratuity is something given voluntarily in return for or in anticipation of a favor or service. Based on this definition, few officers would disagree it is wrong to take a gift knowing the giver expects a "professional favor." However, it is sometimes difficult to determine whether the giver expects something in return. Many people who give gratuities state that they want nothing in return, but in reality they do expect special treatment regarding police services.

There is a big difference between receiving a free cup of coffee from a restaurant and accepting a free set of tires from a local car dealer. Some officers recognize the term "open account" as another way of saying a "payoff." Officers who tell merchants to put merchandise or services on their accounts rarely have intentions of paying for them—they really should not be cops.

It is difficult to be certain what people are thinking when they give gratuities. The only way to be certain that citizens expect no special treatment or favors in return for police services is to not accept gratuities to begin with. This may not be as easy as it sounds if you work within a department where officers routinely take gratuities. Do not give in to peer pressure; stand up for what you know is right.

Perjury

Is it ever ethical for an officer to lie? In many instances deception is justified. Common examples include decoy cops, undercover investigations, or surveillance. In these situations deception is justifiable. In other circumstances it is not, as explained in the following paragraphs.

Can police commit crimes or infractions in order to further law enforcement? Yes, such as when officers speed to catch a speeding traffic violator. Unfortunately, though, some officers commit more serious infractions by claiming the end justifies the means. For example, an officer may exaggerate an incident to establish probable cause for an arrest. A detective may distort circumstances of violations so that he'll have a high clearance rate. A motorcycle officer may intentionally change details of traffic citations in order to issue a certain number during the month.

It is not always easy, but we should all have the courage to stand up for what is fair, right, and honest—it takes a lot of guts. The fact is, once cops distort the truth for any self-serving reason, they lose self-respect. Lying becomes easier with time, and soon intentionally forgetting details or exaggerating facts may be part of sworn testimony during depositions or trial. Such cops do not honor their sworn oath or their badge.

Ticket Fixing

The fixing of tickets has been practiced for decades. It generally means doing something that permits a traffic offender to escape lawful process; usually canceling a ticket. Sometimes friends ask officers to fix tickets, or motorists offer gratuities to officers to reduce or cancel traffic offenses. At other times officers are requested by someone in the department to fix tickets in order to take care of influential or wealthy citizens who have committed offenses.

It is wrong for anyone to be above the law because of wealth or contacts. Law enforcement has a long way to go to overcome the stigma of ticket fixing.

Blind Loyalty, Code of Silence

At the moment a rookie is sworn in, the bond of fellowship begins. Initially, all veteran officers are role models for rookies. New officers look to them for approval, advice, and guidance, and often friendships form. Camaraderie and esprit de corps are stronger than in most occupations.

As the years go by, the phrase "taking care of your own" assumes a special meaning. Cops on the street watch out for each other. They have to be able to trust each other as partners. Loyalty to each other is crucial, and it would be difficult to survive without it.

Bad cops think other officers will not speak up about their corruption. They do not care about law enforcement or fellow officers—only themselves. We know that there are 700,000 officers in the nation. Only about three hundred are decertified each year, less than half of one percent. Perhaps most professions cannot boast of such low numbers. Your job in protecting yourself and the profession is to inform superiors when officers steal or commit perjury. To keep silent would end your career.

The fact is, if you are going to be an officer and good or even great at what you do, have the guts to stand up for what is right. Remain loyal to the principles for which other cops have died. Officers whose weak character allows them to become dishonest and deceitful have no place on the force.

Internal Politics

Disenchantment and a lack of comradeship are common within departments where people place blame or become defensive when things go wrong. Yet it is virtually impossible to develop a strong, healthy character without passing through and learning from difficult times. Learn from difficult situations, correct whatever caused them to happen, and move on to better times. Remember, those who blame others often are to blame themselves.

To develop and maintain a good public image takes a never-ending effort. No agency can afford the time and effort spent bickering and fighting among its own people—far too many problems exist on the street that must be taken care of.

Defending the Badge

America is starting to undergo a rebirth of concern for ethics and integrity, especially in law enforcement. Insider Wall Street trading scandals, Irangate, impeachments, Watergate, sex scandals in the White House, and the arrest of public officials have generated attention to dishonest and unethical practices. Fortunately, even the business world has realized that conscious ethical training has a new tone of morality.

The business community has found that the effort is both worthwhile and efficient. As an example, the 1999 CEO of the year, Herb Kelleran of Southwest Airlines, says that his leadership philosophy is simple: Treat all people with dignity and respect and follow the Golden Rule, and you will be repaid a thousand times over.[10]

The struggle against ethical abuse is ongoing. Keep in mind that a single act of corruption can destroy the reputation of an entire department for decades. In order to gain the respect of those you serve, make a commitment to be an ethical, conscientious cop.

Life isn't always fair. There will be hard times in both your personal and professional lives. It's all part of being human. Accepting disappointment as merely a temporary setback in a long line of accomplishments will allow you to have a much more satisfying, rich life.

No one can control everything that is going to happen on the street, in your community, or in your own department. You can, however, control yourself; for example:

- Make the most of yourself, rather than spending time trying to blame or change others.
- Live for the future, rather than spending time regretting or living in the past.
- Strive to become the best person you can be, rather than developing a habit of criticizing others.
- Discover your own shortcomings and improve yourself, rather than becoming a burden by nagging or arguing.
- Allow other individuals to be human and in control of their own destinies.
- Do not judge others.

Pride, loyalty, and high ethical standards are honorable. Likewise, integrity and professionalism ingrained throughout an agency cause the entire organization to develop a healthy, positive tone. Camaraderie between officers will reinforce the guts that protect the badge and the force from corruption. Law enforcement and every cop in America has respect to earn and a reputation to protect.

END NOTES

1. John Dorschner, "The Dark Side of the Force," The Miami Herald Sunday Magazine, *The Miami Herald,* March 8, 1987, pp. 1–22.
2. Ibid. pp. 1–22.
3. Ibid. pp. 1–22.
4. Lawrence J. Dempsey, "The Knapp Commission and You," *The Police Chief Magazine,* IACP, Gaithersburg, MD, November 1972, p. 21.
5. Ibid. p. 24.
6. Ibid. p. 25.
7. Neal E. Trautman, "The National Law Enforcement Officer Disciplinary Research Project," The National Institute of Ethics, 1998, pp. 1–10.
8. William B. Melnicorr and Jan Menning, "Elements of Police Supervision," Glenco Press, Beverly Hills, CA, 1969, p. 78.
9. International Association of Chiefs of Police, Ad Hoc Ethics Training Subcommittee, Subcommittee Report and Recommendations, 1996.
10. ABC television, excerpt from the 20/20 broadcast of June 13, 1997.

CHAPTER 3

CHARACTER OF GREAT COPS

"The best cops I have known were the ones that didn't change much when they entered the profession. They knew who they were and what they were about long before they put on the badge. In a line of work where someone will question or criticize nearly everything you do, having a firmly anchored self-image is essential. I also look at what sorts of things the person did before they decided to be cops. Did they choose to serve their country in the armed forces? Did they get involved in the community by doing volunteer work? The person that appears to have awakened one morning and decided that they want a career of public service, with no prior inclination to that end, may have other, more self-serving, reasons for wanting to be a police officer."

Timothy M. Dees
Law Enforcement Trainer and Author

The primary reason people apply to be cops is to help others. Cops face and accept frustrations and adversities so that others may be safe. Veteran officers know that years of working the street will probably make these same officers somewhat callous. However, underneath the exterior hard shell beats a heart that is still caring and dedicated, though no longer idealistic.

Why would officers leave their families each day to spend the next eight hours dealing with grief, frustration, and crisis? Why would they want to earn a living dealing with sleazy people, those who have been brutally victimized, or people who have enjoyed being brutal? It takes strong characteristics to rebound from this daily emotional stress. Great cops possess several characteristics: tenacity, sensitivity, maturity, intelligence, physical strength, and more.

Great cops might be born with the right characteristics. They also may have been raised in a way that instills them with the character needed to withstand the frustrations and temptations they face as officers. Great cops can also be just ordinary people who rise above the temptations and frustrations in order to face the responsibilities of a noble, worthy cause.

Individuals applying to become officers learn that the process is rigid. Applicants must complete several hiring procedures: written entrance exam, oral interviews, physical fitness test, written psychological exam, oral exam, and a medical exam with drug screening. Few jobs require applicants to endure such scrutiny.

Professional law enforcement agencies have a thorough hiring process to determine who is suitable as officers. Although the employment examinations are intended to ensure that hiring standards are met, they cannot judge individuals' integrity or their quality.

Many of the things that make great cops are not measured by preemployment processing. Interviews and assessment centers are the best means to identify the qualities of courage, desire, enthusiasm, integrity, caring, and comradeship—all part of "the right stuff." Intuition as well, besides background investigations, can be the best way to evaluate these qualities in the final analysis.

ENTHUSIASM

Some people seem to possess a trait that brings out the best in both themselves and others: enthusiasm. It can move people through hard times and generate the initiative, drive, and excitement needed to overcome adversity. Few great achievements or successes can be accomplished without enthusiasm.

At times, years of verbal abuse and frustration can create a degree of cynicism and pessimism within officers. Without realizing it, they may become negative and critical of virtually everything. However, no team functions at its optimum level if its members think negatively. Without initiative and enthusiasm, few great achievements can occur. Fortunately, enthusiasm is contagious and it costs nothing. Its rewards, though, are great.

COURAGE

Courage means putting your fears aside and getting the job done, which is what officers across the nation do daily. Usually they receive little or no publicity but feel that their actions are part of the job.

Most dictionaries define courage as the ability to meet danger or difficulty with unwavering determination and bravery. Many people often envision courage with shoot-outs, fights, or some form of valor or violence. For officers in America, this vision of courage certainly is real. Senior officers can recall when bravery was necessary.

Sometimes aspiring officers feel attracted to the excitement of working the street. They can, however, generate their own crises by overreacting during emotion-filled situations, thereby becoming part of the problem instead of the solution. Some died needlessly and were hailed as courageous, but created their own deadly circumstances because of overzealousness. Others were insecure and thought they had to prove themselves.

Never mistake courage for panic, the John Wayne syndrome, or simply the inability to remain calm, often referred to as "tombstone courage." This tendency is usually seen in inexperienced or immature officers, so experience can provide the necessary wisdom. Law enforcement has an obligation to be courageous if a situation demands it, but actions should always be balanced with intelligence and discretion.

Most acts of courage are thought to be heroic, but it takes courage to carry out many of the tasks on a tour of duty. Most, if not all, veterans have been afraid of the street at one time or another. There is nothing wrong with controlled fear, as long as it can keep you alert and alive. In fact, in some situations fear is beneficial, as long as you are able to control it.

Imagine patrolling a high-crime area as a single-officer unit. It is 0200 hours on a midnight shift as you enter the parking lot of a bar where there are frequent fights, knifings, and shootings. Your city's ordinance prohibits having open alcohol containers outside of bars or liquor stores. However, while driving slowly through the parking lot, you discover four or five local adults and two are drinking beer.

Although the drinking adults know that their actions are against the law, they also know that some officers look the other way rather than confront them. Some citizens think that glares and stares intimidate officers, but officers who submit to this type of intimidation make it difficult for both themselves and fellow officers to uphold the law.

Discretion is always appropriate, but it takes guts to confront several lawbreakers. Officers must make confrontations wisely. Single-officer units should request another officer for assistance and backup. Once the backup arrives, the confrontation can begin. If another officer is not available, the single-unit officer should "check out" with the dispatcher before confronting the suspects. Use survival tactics, such as proper stance, voice commands, positioning, and use of light.

It is uncommon that cops need to react heroically. When it is necessary, best effort is essential as is common sense and the use of skills developed through officer training. Use discretion, yet still enforce the law; it is foolish to instigate or escalate confrontation. Instead, react with wisdom, skills learned through training, and strong physical ability.

POSITIVE OUTLOOK

Cynicism in any profession is detrimental to both a career and personal life. It destroys a healthy and positive attitude. Most people get out of life what their perspectives allow. In other words, if they expect the best and work hard to achieve it, they usually will do so. On the other hand, if they are negative and suspicious, the self-fulfilling prophecy will come true. When things go wrong, it helps to view them as temporary setbacks in a long line of accomplishments.

Police sometimes deal with the worst side of humanity. Some officers, as in any profession, become hardened without realizing it. Others may be constant complainers. They may be critical of or dislike their organization, fellow officers, citizens, or themselves. Their personal life is usually equally depressing.

Some cops claim that officers should be cynical because good street cops are suspicious by nature. You can still be positive and optimistic and be a great cop. Attitude is vital and a positive one helps people—and officers—deal with life and its problems better and more wisely. Negative, unhealthy attitudes cloud proper judgment.

Leading authorities on performance and leadership agree: the formula for personal success is 85 percent attitude and 15 percent skill. Setbacks and bad times are sure to happen to all of us. It is best to move on enthusiastically when they happen, rather than spending energy and time worrying, complaining, or criticizing. Negative, apathetic people lead dull, uneventful lives. Those who are successful, happy, and satisfied with their lives tend to be enthusiastic. Positive thinking is an acquired trait. People aren't born with an optimistic attitude; they have to acquire it. Enjoy the simple moments of life: laughing, caring, smiling, loving, and sharing. Above all, persevere with positive thinking and reasoning.

DISCIPLINE

Law enforcement is a tough, demanding profession. Almost every aspect of it requires discipline. For example, it takes discipline to stay in good physical condition, know the facts of a case before testifying, conduct aggressive patrol on midnight shift, deal with the frustration of the judicial system, or ignore the sarcasm of citizens.

Discipline means hard work, but it offers many satisfying rewards. Disciplined individuals are more satisfied with themselves. They understand and accept self-control and effort because they benefit from the healthy self-image and promising future that results from them.

Self-esteem is essential to personal satisfaction. It also helps you feel good about your job and work. If you lack discipline, make a sincere commitment to improve. Do not procrastinate; as you begin to accomplish more, accomplishments will abound, and discipline will become habit-forming.

CARING ATTITUDE

Most people respect the police and even admire them. This silent majority understands how tough law enforcement is and respects officers for doing their best. The truth is, though, most citizens do not understand cops or the job that they do. It is sometimes difficult for them to get to know officers, but educating them may change their attitudes.

Although it is hard for someone with no police experience to understand, every veteran cop knows that a lot of officers feel most comfortable when they socialize only with other cops. However, this is not a healthy situation.

Cops have a tendency to be reserved in nature, even among other officers. This may be because as the years pass, they develop callousness as a defense against the continual suffering, pain, and frustration they witness. Yet, beneath that hard exterior usually beats the heart of someone who really cares.

Officers must remain sensitive toward others and avoid self-centeredness in their hectic, stressful lives. They must remember that despite the crises and despair they face on the street or behind the desk, they have an honorable purpose and profession.

If they should ever stop caring for those they serve, they need to seek assistance. It would be helpful also to reevaluate their motives, beliefs, and important aspects of the job. Sincerity is extremely important to good cops, and caring comes from appreciating what's really important. Robert Hastings summarized it best in his essay "*The Station.*"

THE STATION

By Robert J. Hastings

Tucked away in our subconscious is an idyllic vision. We see ourselves on a long trip that spans the continent. We are traveling by train. Out the windows we drink in the passing scene of cars on nearby highways, of children waving at a crossing, of cattle grazing on a distant hillside, of smoke pouring from a power plant, of row upon row of corn and wheat, of flatlands and valleys, of mountains and rolling hillsides, of city skylines and village halls.

But uppermost in our minds is the final destination. On a certain day at a certain hour we will pull into the station. Bands will be playing and flags waving. Once we get there so many wonderful dreams will come true and the pieces of our lives will fit together like a completed jigsaw puzzle. How restlessly we pace the aisles, damning the minutes for loitering—waiting, waiting, waiting for the station."When we reach the station, that will be it!" we cry. "When I'm 18 . . ." "When I buy a new 450 SL Mercedes Benz . . . !" "When I put the last kid through college . . ." "When I get married . . ." "When I have paid off the mortgage . . ." "When I get a promotion . . ." "When I reach the age of retirement, I shall live happily ever after!" Sooner or later we must realize there is no station, no one place to arrive at once and for all. The true joy of life is the trip. The station is only a dream. It constantly outdistances us.

"Relish the moment" is a good motto, especially when coupled with Psalm 118:24: "This is the day which the Lord hath made; we will rejoice and be glad in it." It isn't the burdens of today that drive men mad. It is the regrets over yesterday and the fear of tomorrow. Regret and fear are twin thieves, who rob us of today.

So stop pacing the aisles and counting the miles. Instead, climb more mountains, eat more ice cream, go barefoot more often, swim more rivers, watch more sunsets, laugh more, cry less. Life must be lived as we go along. The station will come soon enough.[1]

SELF-CONTROL

In crises officers can never lose self-control. When others are shouting, screaming, and panicking, the cops are responsible for maintaining the calmness and logic necessary to handle situations. Panic is a luxury they cannot afford. Therefore, individuals who have trouble controlling their emotions should choose another profession.

Staying calm is easier said than done. Some are more adept at it than others, but there is no magic formula for controlling anger and maintaining calmness. Officers should remember to be professional, or their careers may be short-lived if emotion overrules logic. They may raise their voices, clench their fists, tense their muscles, and make their bodies rigid. They are being rough on their bodies and probably making fools of themselves.

To prevent people from "getting under your skin," do not let them upset you in the first place. Again, this is easier said than done. If it begins, keep your hands from clenching and lower your voice. If you can, bring your voice down to a whisper, because it will be difficult for people to argue with you when you speak softly. Control your emotions—deliberately and conscientiously. After the crisis if you are frustrated, talk about it with someone who knows how to listen or manage emotions well.

The days are gone when cops felt they had to be superhuman. Do not suppress your emotions following an incident because they will build up inside you. Use your agency's counselors or support groups, or talk to a fellow officer, a friend, or your spouse.

Cops see and feel frustration and futility, and it is not unusual for them to need help in maintaining sound mental health. A cop should have the ability to adapt to different situations by losing the feelings of hopelessness and unfamiliarity. If you need assistance and your agency does not have a counseling program, argue for one. More than likely you will have fellow officers who must understand the consequences of suppressing emotions over an extended period of time and also need counseling to work through them.

Traits of Great Officers

- The ability to accept people the way they are and then deal with them in their own frame of reference.
- The ability to think independently and make decisions based on facts, law, and what you know to be true and right.
- Physical fitness, primarily for personal well-being and stress reduction. It also helps in job performance.
- The drive to do your best and want to make a difference in some way, even if only a few people can be helped.
- Flexibility, which allows you to get the job done and still deal with politics. It also allows you to adjust to different jobs, different supervisors, and different working conditions.

WISDOM

Wisdom, which can be defined as a high degree of knowledge, is something every officer needs. Officers are called upon to be a minister in times of grief, a tireless documenter of facts, individuals who can make split-second legal decisions, and diplomats who can talk with a transient one moment and a politician the next. They must know about numerous state statutes, local ordinances, departmental rules and regulations, and standard operating procedures. Further, a cop's wisdom should be balanced with empathy for the unfortunate and unyielding firmness for the intimidating.

Paul Harvey, the highly respected commentator, has always been a friend of law enforcement. He summarized officers' duty and wisdom one day during a broadcast:

> He, of all men, is at once the most needed and the most unwanted. He is a strangely nameless creature who is "sir" to his face and m—— f—— behind his back. He must be such a diplomat that he can settle differences between individuals so that each would think he won. But if the policeman is neat, he is conceited; if he is careless, he is a bum. If he is pleasant, he is a flirt; if he is not, he's a grouch. He must be first to an accident, he must be able to start breathing, stop bleeding, tie splints, and above all, ensure the victim goes home without a limp or expect to be sued.
>
> The police officer must be able to whip someone twice his size and half his age without damaging his uniform and without being brutal. If you hit him, he's a coward. If he hits you, he is a bully. A police officer must know everything and not tell. He must know where all the sin is and not partake. The police must, from a single human hair, be able to describe the crime, the weapon, and the criminal and tell you where the criminal is hiding. But if he catches the criminal, he is lucky, and if he doesn't, he is a dunce. If he gets promoted, he has political pull; if he doesn't, he is a dullard.
>
> The policeman must chase bum leads to a dead end, conduct a stakeout ten nights to tag one witness who saw it happen but refuses to remember. He runs through files and writes reports until his head aches to build a case against some villain who will get dealt out by a shameless attorney or judge who isn't honorable.

In part, wisdom is nothing more than common sense. It is the ability to be logical and practical, although it takes more than common sense to be wise. It requires a lot of knowledge, tenacity to learn state statutes, city ordinances, and departmental regulations. With few exceptions, knowledge is acquired only through hard work and dedication.

Every street officer deals with people who are upset. Officers must be prepared for the unexpected and always think "survival" in their minds. They cannot interpret citizens' anger as a personal offense. Great cops can distinguish between someone venting their frustration and an individual having committed a crime and using verbal abuse. The individual must be arrested without anger and excessive force.

Do not take things personally, just treat others as you would like to be treated. Remember, maybe they are good people who are just frustrated, angry, and need someone to listen. How you listen and respond may determine if someone gets hurt or if the situation is defused and resolved. Be logical and understanding, and let them tell their side of the situation. Psychologists confirm that it is very therapeutic to allow someone to vent anger, so you are doing everyone involved a great service.

This is not always easy; sometimes it requires an enormous amount of control and patience. The following story illustrates the virtue of patience.

> One day a young boy was walking by himself through the woods. Curious and adventurous, he had to stop and look at anything and everything. He suddenly discovered a cocoon on the branch of a tree.
>
> As fate would have it, a butterfly was making a hole, preparing to break out of the cocoon. The boy's impatience caused him to help the butterfly out of the hole, and the miracle of life took place before his eyes: the case broke open and the butterfly slowly emerged.
>
> To the young boy's horror, he realized the butterfly's wings were still folded back and unable to open. The butterfly wrenched with all of its strength, but was unable to unfold them. So, the boy, feeling helpless and frustrated, unfolded the butterfly's wings. After a few moments, the beautiful butterfly died in the boy's hands. The boy was filled with guilt and realized too late that the butterfly's wings should have unfolded naturally. One of nature's most beautiful things had died in the palm of his hands after a brief, yet desperate struggle, and the young boy's impatience had caused its tragic death. Nature had been interrupted, and the boy's lack of control had taught him a horrible, but important lesson.
>
> It was a long and tiring walk out of the woods for the boy; the tiny butterfly weighed heavily on his conscience. He had learned a lesson that he would never forget, that patience and control are crucial. For cops, the lack of patience and control—and wisdom—can end in human tragedy.

One of the most difficult lessons to learn is also one of the most valuable when handling people who have lost their temper: controlling your own emotions is absolutely vital. Some officers become defensive and feel the need to "strike back" or take revenge; the answer to this temptation is to not allow yourself to become angry. It cannot even be an option; it's an essential commitment. Controlling your emotions will allow you to work more effectively and professionally and make sound decisions.

HARD WORK AND DEDICATION

While there are exceptions, most worthwhile things in life come with a price. In one way or another the price usually is hard work and dedication. Herschel Walker, one of football's greatest athletes, summed up his philosophy on life when he said, "The Lord places many challenges ahead of us, and the only way to meet them is head on." Herschel was probably saying there is no free ride in life. Hard work and determination become a way of life for those who accomplish the most. In doing so, they feel good about themselves, and their self-esteem and self-worth flourish.

Police work has been described as 95 percent boredom and 5 percent pure hell. For most agencies that is an accurate description. Even the mundane, boring times can involve hard work. When calls are backed up and an agency is shorthanded, a shift can be challenging, to say the least.

Whether or not you are a street cop, you are probably overworked and understaffed. Detectives sometimes do paperwork most of the day; having time to go out on the street can be a luxury. With 50 to 70 cases per month, an investigator may sometimes only interview witnesses or victims and write reports.

Other divisions and units are just as overworked. Hard-working officers and their friends and acquaintances have many rewards. They generally enjoy working, facing challenges, and achieving personal goals and have a zest for life. Their days are filled with hard work, yet they like it that way because achievers thrive on challenges and accomplishments. Usually, too, they are optimistic about the future. Therefore, it is easy to see that hard-working dedicated optimists will make outstanding cops. On the other hand, people who are somewhat lazy or lack initiative would rather get out of work than face it head on. These people generally are not happy with other aspects of their life and tend to complain, be pessimistic, and feel dissatisfied with everything.

SUPPORT, TRAINING, AND DEVELOPMENT

Emotional trauma and frustration can jeopardize officers' mental health. As a result, marital or personal problems can be magnified as pressure continues to build within the individuals. Although law enforcement has typically not been good at taking care of its officers, many of them would benefit from professional counseling at some point during their career. It could result in more effective, productive work, and happier, healthier people. Communities owe their law enforcement forces this assistance and support.

Agencies, in addition to professional counseling services, should provide adequate training for their officers. Historically, many have offered inappropriate or little training. The last decade has seen vast improvements, but some departments still fail to train personnel properly. Commitment to train must begin at the top, with the chief, sheriff, director, or superintendent, although this level does not always appreciate the value of and need for continual, effective training.

Individuals assigned the responsibility of training may not have been taught the most effective way to develop and deliver it. To do so, a thorough needs assessment and job-task analysis are essential as well as the development of realistic learning goals, objectives, lesson plans, and pre- and posttests. Effective training will help boost morale, enhance necessary skills, and save officers' lives.

Every organization should realize that its greatest resource is its employees. Yet, many American corporations and acclaimed management textbooks agree that most organizations fail to provide sufficient employee development. In other words, they do not make the most of what employees have to offer. The skills, abilities, and potential of employees should be discovered, developed, and used.

Human resource programs can simply determine employees' skills and abilities, record them in a computer database for easy reference, and then determine ways to use them to everyone's advantage. Not only will this offer tremendous benefit to an organization, but employees will feel a new sense of self-worth by knowing that their administration appreciates their talents and abilities.

EDUCATION

The Police Executive Research Forum (PERF) is one of the most respected law enforcement organizations, recognized for its leadership and guidance. PERF conducted an extensive nationwide study in 1988 to determine the status of contemporary law enforcement education. Six hundred ninety-nine agencies were surveyed, and 67 percent responded. The findings are enlightening.

Survey Findings

Educational Policies for Sworn Officers

- 62% of the responding agencies have at least one formal policy in support of officer higher education. Most departments have more than one such policy. These policies serve as a measure of demonstrated support for higher education.
- 58% of agencies with educational support policies require that the coursework be job related.
- "Job related" may include various subject areas: 49% indicated a preference for criminal justice majors, 46% indicated no preference for the major.
- Related majors, indicated from comments and site visits, include business administration, sociology, psychology, and social sciences. Some also indicated a preference for general liberal arts degrees.
- Those indicating a preference for criminal justice graduates did so because of their enhanced knowledge of the entire criminal justice system and issues in policing.
- Supporters of the liberal arts degrees argue that such college preparation gives individuals a wider range of tolerance and decision-making resources.

- Only 25% of the agencies with educational policies required that the coursework be part of a degree program. This is somewhat surprising, because without requiring coursework to follow a degree plan would mean personnel could take a variety of unrelated courses.
- Individuals in support of a college degree as a symbol of a "professional" would find this approach wasteful of resources and lacking direction for professional achievement. Conversely, you could also argue that it takes a variety of skills to be an effective police officer.
- 61% of the responding agencies had a collective bargaining agreement; 54% of these had educational incentive provisions as part of the contract.
- Based on comments and information from site visits, it is reasonable to assume that the incentive provisions were, for the most part, proposed by police officer associations (POAs), not management. POAs appear to give initial support for educational requirements for employment, believing this may lead to increased salaries.

Educational Policies for Police Civilian Employees

- 51% of responding agencies provided some form of educational support for civilians.
- Most of these policies are local government programs rather than exclusively police department policies.
- The most common civilian educational policy is tuition assistance for job-related courses.

Requirements for Entry into Police Service

- 14% of the law enforcement agencies had a formal college requirement for employment, a somewhat lower level than anticipated.
- Municipal departments are more likely to have a college requirement than consolidated sheriffs or state agencies.
- The most common higher education requirement was a 2-year, 60-hour associate degree.
- The basis for the number of college credits required for employment seems to be an intuitive decision based loosely on the recommendations of the President's Commission and National Advisory Commission, the ability to recruit persons with college credits, and the effects of the college requirement on potential minority applicants.
- The practice of police employment indicates that the majority of police departments have an informal policy to at least give preference to those with college training
- Reasons why agencies do not establish a formal education requirement for employment include:
    - A presumed discriminatory effect of the higher education requirement.
    - The fear of a discrimination lawsuit for an educational requirement.
    - Establishment of the educational requirement requires validation, and agencies are unsure how to validate higher education as a bona fide occupational qualification.

  - Agencies fear that some good police candidates may be lost as a result of the requirement.
  - Agencies and parent jurisdictions fear that POAs may seek increases in salaries and benefits if a higher education requirement is established.
  - With higher education there must be more aggressive—hence, more expensive—recruitment efforts.
  - Agencies fear that they may be unable to fill academy classes.
  - Administrators feel that higher education is a necessity or at least a benefit, yet they are not totally convinced.
  - Interference by POAs and/or city personnel departments makes the establishment of an educational requirement more difficult.

Promotional Policies

- 75% indicated they had no policy or practice requiring college education for promotion.
- 8% had a written policy requiring some college hours for promotion, however, the number of required credits varied.
- 3% reported informal policies requiring college hours for promotion. Survey comments indicated that even more than the 3% may have the informal requirement, however, they did not report it.
- 4% provided early promotion eligibility for persons with college hours.
- 82% did not require college credits for promotion but recognized college education as an important element in promotion decisions.

Recruiting

- 54% directed recruiting equally at college students and the general public.
- 7% primarily directed recruiting efforts toward minorities and women, regardless of education.
- 27% had either a full-or part-time education liaison officer.
- The most prevalent student-oriented program (59%) was a Scout Explorer Post associated with the police department. Site visit information indicated that the Explorer Posts were educational, but relatively few explorers pursued a law enforcement career with the agency.
- Other significant student-oriented programs existed indicating that many departments have policies and practices to establish liaisons with college-educated groups.

Important Factors Relating to Colleges

- The most important factors for an officer selecting a college to attend were location (82%), cost (72%), and degree offerings (72%).
- 70% of police chief executives indicated that they had been consulted by area colleges/universities at least a few times on curricular and/or research matters.

Advantages and Disadvantages of College Education for Police

- Overall, the results suggest that police executives find more advantages to college-educated officers than disadvantages.

- Among the more salient indicators reported by the respondents were that college educated officers:

  communicate better with the public,
  write better reports,
  perform more effectively,
  receive fewer citizen complaints,
  show more initiative in performing police tasks,
  are more professional,
  use discretion wisely,
  are more likely to be promoted,
  are better decision makers,
  show more sensitivity to racial and ethnic groups, and
  have fewer disciplinary problems.

Education, Race, and Sex

- The average educational level for police in 1967 was 12.4 years, barely more than a high school diploma.
- Current average educational level is well into the sophomore year in college; given the time involved for social change to occur and the large number of police officers involved, this increase is qualitatively significant.
- The variance among the educational levels of the racial/ethnic groups is small.

Table 3.1 gives additional information on education by race and ethnicity. Table 3.2 provides similar information by sex of officer.

TABLE 3.1
EDUCATIONAL LEVELS OF OFFICERS BY RACE/ETHNICITY

Officers	*Average Level of Education*	*No College*	*Some Undergraduate Work*	*Graduate Degree*
Blacks	14 years	28%	63%	9%
Hispanics	13 years	27%	68%	5%
Whites	14 years	34%	62%	4%
Other Race/ Ethnicity	14 years	19%	73%	8%

- The consistency of no college and some undergraduate work between the races is particularly striking. The data indicate that minority group members with higher education can and are being effectively employed by law enforcement agencies.
- College-educated minority group members can be effectively recruited for law enforcement.
- There may not be a need to establish differential educational criteria for minority group members to meet affirmative action goals.

TABLE 3.2
EDUCATIONAL LEVELS OF OFFICERS BY SEX

Sex	*Average Level of Education*	*No College*	*Some Undergraduate Work*	*Graduate Degree*
Males	14 years	35%	62%	3%
Females	15 years	24%	46%	30%

Implementing Educational Requirements

All cities visited had policies and practices in support of higher education. The sites visited were: San Diego, San Jose, and Sacramento, California; Tulsa, Oklahoma; New York City, New York; Kansas City, Missouri; and Largo, Florida. These cities revealed a number of implementation issues:

- Educational requirements for entry and promotion can be effectively implemented.
- Educational requirements are most successful when supported by a plan for implementation and programming.
- Agencies with higher education requirements for employment tend to have nontraditional recruitment programs.
- Innovative policy alternatives to facilitate the attendance of college were noted in most agencies.
- Organizational commitment to the educational program was fundamental to the program's success.
- Ongoing dialogue with police officer associations prior to implementation of any educational criteria facilitates program development.
- Educational requirements are best implemented as a matter of policy, not informal mechanisms.
- Ongoing communication between the police department and local colleges and universities (most notably through criminal justice programs) enhances the ability to implement higher education policies; this was a factor strongly evident at all cities visited.
- In those agencies with a higher education standard, educational support policies were evident (e.g., policies with incentive pay, tuition assistance, permitting on-duty class attendance, permitting officers to exchange days off and/or shift assignment to facilitate class attendance).

END NOTES

1. Robert J. Hastings, *A Penny's Worth of Minced Ham: Another Look at the Great Depression,* Shawnee Books, Southern University Press, 1986, pp. 90-1.

CHAPTER 4

COPS AND THE JUDICIAL SYSTEM

"A great cop is an individual committed to service with the courage to do what is right—and understands that our laws were developed to act as tools to solve problems, not to be used as clubs to abuse communities."

Louis M. Dekmar, Chief
La Grange Police Department
La Grange, GA

The dignity of crime victims is often ignored by the criminal justice system, as repeat offenders are set free over and over again. It seems that the system is filled with stagnation, injustice, and inequity.

The frustration that is felt by citizens with the justice process is not new, nor is it a burden known only to America. Injustice occurs in all nations, but some nations deal with judicial concerns more effectively than others.

Most cops feel that the American criminal justice system has its good points. Yet they, too, face frustrations with it. Sometimes criminals are released before officers have completed the related arrest reports. Plea bargaining, early paroles, and a seemingly endless appeal system are more than irritating. Legal technicalities, liberal judges, and a process that appears to forget victims frustrates and complicates the efforts of the police.

Officers play one of the most crucial roles in the judicial system, so they must understand how the system works. Unfortunately, many officers pass through their career and never watch a felony trial in its entirety. Some are never exposed to the problems and complexities of other phases within the system. This lack of understanding can lead to further agitation and it certainly perpetuates more problems. Similarly, when officers have no working knowledge of legal terms and their meanings, they can experience disastrous results to their cases and their credibility. (A legal glossary is provided near the end of the chapter to help you become familiar with important vocabulary.)

THE LAW

The criminal justice system is America's means of social control. It is intended to protect society by apprehending, convicting, and rehabilitating offenders of criminal law. Learning more about the process is interesting, revealing, and of course, frustrating. Officers should know how their actions affect other parts of the system.

Millions of people are crime victims each year, and crime directly influences the quality of American life. The criminal justice system, therefore, is held accountable for controlling it. Since crime is directly dependent upon and affected by a variety of social ills, the challenge is monumental.

The justice system is enormously complex, but it is still only one part of the process of law. To appreciate the process, we must understand how American criminal law developed. Laws are the formal means of control within a society. They afford society with protection and safeguard against physical harm and infringement of personal rights. Life would be reduced to survival of the fittest without laws.

Laws are one form of social control. Through tradition, cultural habits, moral standards, peer pressure, and ever-changing fads, people can also conform to behavior. In one respect, the law is much different than other means, because negative consequences have been established for violating them. The justice system regulates the nature, method, and extent of the consequences.

Primitive societies depend on social customs, tradition, and rules for social regulation. Rules are similar to laws, but lack the judicial complexity associated with laws. The judicial process and its associated subsystems set laws apart from rules. As an example, rules may be enforced by family members or private organizations. Laws, however, are enforced by government and its political divisions.

American law is derived from English law. As such, it is also known as Anglo-American criminal justice. Once again, the difference between it and other methods of social control was its complicated decision-making procedures.

More than two hundred years ago America's leaders met to discuss, debate, and develop an everlasting guideline for our country. This document, the Constitution of the United States, guarantees the freedom and rights of all Americans. Police officers dedicate their lives in order to uphold the vision our forefathers encompassed.

Law and its enforcement serves many purposes. Some of the practical benefits include:

- Settling disputes
- Deterring crime
- Punishing criminal activity
- Preserving order
- Assigning authority among governmental and social agencies.[1]

Throughout the years, patrol officers—the front line of defense—have faced an enormous responsibility: to protect citizens while ensuring their guaranteed rights. To protect the communities officers are sworn to serve is not always easy. America is a violent society; each year millions of Americans are victims of crime. The Constitution can guarantee a regulation of commerce, taxation of citizens, the interrelationship of federal and state governments, and the rights of those accused of crimes, but it can only attempt to govern human behavior.

The causes of society's ills are complex and multifaceted. The police have fought hard and long to combat crime, usually going far beyond what could be expected, especially given the hardships and tragedies they must face. Law enforcement officers are sworn to uphold and enforce the Constitution and all the rights and liberties it guarantees. Thousands of officers have given their lives in the daily struggle for protection and justice—the ultimate sacrifice so that others may live with liberty and justice, law, and order.

FIGURE 4.1

Goals and Purposes of Criminal Law and the Criminal Justice System

1. To discourage and deter people from committing crimes.
2. To protect society from dangerous and harmful people.
3. To punish people who have committed crime.
4. To rehabilitate and reform people who have committed crimes.

Source: Thomas J. Gardener, Criminal Law, West Publishing Co., New York, 1988; p. 20.

Historical Perspectives

From its initial efforts at colonization, America has felt a commitment to protect the individual rights of citizens from the abuse or misuse of government. This is partially due to the fact that the United States was founded as a result of Britain's abuse of early colonists. The original Constitution of 1788 and the Bill of Rights resulted from our commitment to protect against abuse.

Delegates from all of the original thirteen states except Rhode Island met on September 17, 1787. Their purpose was to propose a new Constitution to the Continental Congress and ratify the states. The origins of the Constitution and its amendments (four years later) date back to ancient Greece and Rome, and several hundred years of English history helped to further refine attitudes and perspectives. Though the courage and insight of these delegates are obvious, there was considerable debate and disagreement. Several delegates strongly disagreed; in fact, Rhode Island refused to attend or sign the document.

The first ten amendments were proposed by Congress on September 25, 1789. These are now referred to as the Bill of Rights. Ratification of the Bill of Rights was not completed until December 15, 1791. Since this date, it has represented America in its commitment to the importance of citizens' rights and limitations toward government. The Bill of Rights protects the liberties that each of us cherish.

FIGURE 4.2

Development of Criminal Law in the United States

English criminal law as developed in the common law of England by English judges who adopted customs and usages.

First Stage in the American Colonies

- Use of English common law as it came across the Atlantic with the early settlers.

Second Stage in the Independent America States

- Under the Articles of Confederation enactment of common law
- Continued use of other common crimes into statutes and law crimes in either their ordinances by newly created original or modified forms
- American legislative bodies

Third Stage with the Establishment of a Federal Form of Government Under the Present Constitution of the United States

- In 1812, the U.S. Supreme Court held that federal courts had no authority to adopt common law crimes.
- Further definition of crimes by state legislatures with the creation of many new crimes in the past fifty years.
- Diminished use of common law crimes in all states with more than half the states abolishing common law crimes. Therefore, all federal criminal prosecutions "must be sustained by statutory authority."

Source: Thomas J. Gardener, Criminal Law, West Publishing Co., New York, 1988, p. 21.

Types of Laws

Police officers must understand criminal law. Most officers deal with statutory laws, especially those that have been enacted by state legislature. Criminal laws are punitive in nature, and violations result in punishment.

Civil laws are noncriminal. They concern matters like civil disputes, contracts, property litigation, personal grievances, or other disagreements which do not violate a criminal law. Officers must be able to distinguish between situations over which they have authority to take action and those which are civil. They should be careful not to give legal advice in civil situations; rather, they should advise citizens to consult with attorneys.

Protecting citizens' rights while maintaining law and order is difficult. Laws and criminal procedures were developed to maintain the delicate balance that the pursuit of justice requires. As more legal safeguards are implemented to protect individual rights, the more difficult it is to maintain law and order. On the other hand, if legal regulation is decreased the protection of citizens becomes easier, but individual liberty is jeopardized.

Limitations

Laws are affected by many social factors; their development and enforcement is subject to influence and has limitations. Understanding the limitations of laws allows us to deal with them more effectively.

Laws are subject to interpretations. Written laws must be interpreted to be enforced; therefore, some inequality exists. Officers apply their interpretation of statutes to particular situations to decide if an offense has been committed and whether there is sufficient evidence for an arrest. Prosecutors interpret the laws and apply them to cases. Judges interpret "case law" and compare it to circumstances of cases. The appeal process evolves around comparing specific circumstances to the interpretation of law. Legal interpretation is influenced by changes in social attitudes and customs, and there is a natural evolution in interpretation itself.

Enforcement of laws requires facts. Our system of justice demands that the prosecution and defense determine and present facts through testimony. The finding of truth is sometimes hampered by witnesses who lie and by important evidence which cannot be presented to the jury for particular reasons. Apathy, fear, or unavoidable circumstances occasionally prevent the determining of fact. In the pursuit of representing their "side" of a case, attorneys may even exaggerate or mislead jurors. Presenting facts in the furtherance of justice can become complicated and difficult.

Law itself cannot control crime. Making a particular act illegal does not prevent it from occurring. Laws offer protection from unacceptable behavior, yet there will always be individuals who, for an assortment of reasons, commit the acts regardless of legislation. Laws are limited to merely forbidding actions and specifying consequences. Someone somewhere will refuse to conform.

Creation and enforcement of laws are subject to unjust influence. The judicial process is subject to human weaknesses and frailties because it depends on individuals who work within it. Legislators may be unfairly influenced during the legislation process. Police officers, attorneys, and judges sometimes yield to favoritism or corruption. These and other influences limit the ability to enforce the laws.

Laws change slowly. Everyone has heard the phrase "there ought to be a law." Making that wish come true is very time-consuming. Sometimes the need for a law exists, but it goes unheeded until legislation acts. In other instances unfair or outdated laws remain in effect beyond their realistic usefulness. The evolution of law is a neverending process. By its very nature, legislation will lag behind changes in society's or a community's attitudes.[2] As changes in laws and their interpretation occur, they are merely a reflection of how society has changed.

Government Branches and the Law

It is important to understand how the basic rights afforded to us by the Constitution and the Bill of Rights are applied in everyday life. Officers should understand how laws are enforced in order to appreciate the significance of criminal procedure.

The three branches of government are judicial, executive, and legislative. Each is required by the Constitution to protect individual rights. The judicial branch is charged with the greatest role; it has the duty to declare what is law and then interpret it. The courts must decide when laws conflict with the Constitution and should be declared invalid.

Congress is responsible for enacting legislation which guarantees constitutional rights and applies them to specific situations. Like the courts, Congress is charged with an awesome responsibility. Once laws are passed by Congress, the executive branch is responsible for implementing them. This branch creates procedures and regulations pertaining to the administration of law.

Other forms of regulations, such as city or county ordinances, exist, although Americans generally live under the federal and state governments. The power of the federal government is limited only by the Constitution. Federal agencies use their authority for matters such as relationships with foreign governments, disputes among states, and situations of national concern. States, however, concern themselves with matters within their domain.[3]

Every cop in America should understand the United States Constitution. The failure to understand it, especially the Bill of Rights, can increase the probability of violating it. The Bill of Rights are the first ten amendments to the Constitution. They have an enormous influence upon officers' actions, their role in society, and the mission of law enforcement. The amendments were written as safeguards to individual liberty. To violate the amendments most likely means violating the rights and privileges of a defendant. In doing so, particular evidence or an entire offense may be dismissed. The Bill of Rights are presented in Figure 4.3.

FIGURE 4.3

Bill of Rights

Amendments to the Constitution

Amendment 1

Congress shall make no law respecting an establishment of religion, or prohibiting the free exercise thereof; or abridging the freedom of speech or of the press; or the right of the people peaceably to assemble, and to petition the Government for a redress of grievances.

Amendment 2

A well-regulated militia being necessary to the security of a free State, the right of the people to keep and bear arms shall not be infringed.

Amendment 3

No soldier shall, in time of peace, be quartered in any house without the consent of the owner, nor in time of war but in a manner to be prescribed by law.

Amendment 4

The right of the people to be secure in their persons, houses, papers, and effects, against unreasonable searches and seizures, shall not be violated, and no warrants shall be issued but upon probable cause, supported by oath or affirmation, and particularly describing the place to be searched, and the persons or things to be seized.

Amendment 5

No person shall be held to answer for a capital or otherwise infamous crime unless on a presentment or indictment of a Grand Jury, except in case arising in the land or naval forces, or in the militia, when in actual service, in time of war or public danger; nor shall any person be subject for the same offense to be twice put in jeopardy of life or limb; nor shall be compelled in any criminal case to be a witness against himself, nor be deprived of life, liberty, or property, without due process of law; nor shall private property be taken for public use without just compensation.

Amendment 6

In all criminal prosecution, the accused shall enjoy the right to a speedy and public trial, by an impartial jury of the State and district wherein the crime shall have been committed, which district shall have been previously ascertained by law, and to be informed of the nature and cause of the accusation; to be confronted with the witnesses against him; to have compulsory process for obtaining witnesses in his favor, and to have the assistance of counsel for his defense.

continued

Amendment 7

In suits at common law, where the value in controversy shall exceed twenty dollars, the right of trial by jury shall be preserved, and no fact tried by a jury, shall be otherwise reexamined in any court of the United States than according to the rules of the common law.

Amendment 8

Excessive bail shall not be required, nor excessive fines imposed, nor cruel and unusual punishments inflicted.

Amendment 9

The enumeration in the Constitution, of certain rights, shall not be construed to deny or disparage others retained by the people.

Amendment 10

The powers not delegated to the United States by the Constitution, nor prohibited by it to the States, are reserved to the States respectively, or to the people.

Source: From John L. Sullivan, Introduction to Police Science, *McGraw-Hill Book Co., New York, NY, 1971, pp. 161–163, 353–354.*

AMERICAN CRIMINAL JUSTICE

Few see crime like the police. Officers see and deal with the victims and their despair and face the hostility of criminals. They are the individuals people turn to in times of turmoil and confusion; frequently no other individuals can help.

A street officer's job can include times that are challenging, boring, even moments filled with terror, and moments that are rewarding. Throughout all these moments society expects officers to act the way cops do in television programs or in the movies. Society wants a happy ending. Happy endings may be easy to write, but they are not always reality.

Fortunately, the American justice system is a real effort to protect citizens while it helps officers defuse the confrontations, violence, and tragedies. Officers are often referred to by citizens as the thin line between society and the violence that is heard and read about daily. The police are essential to society and everyday life. Without them the justice system of our nation would soon begin to crumble.

Our system of criminal justice has three distinctive components: law enforcement, courts, and corrections. Each has separate responsibilities, yet they are interwoven and dependent upon each other. The effectiveness of each phase directly influences other areas of the system. Within every subunit or major component, an organized, complex sequence of procedures occurs. The process is intended to ensure that individuals' rights are protected. No citizen may be punished until being afforded due process of legal protection.

While the protection of rights is important, the criminal justice system must also safeguard citizens. The apprehension, prosecution, and punishment of criminals is a primary concern. The American justice system intentionally sacrifices part of its effectiveness in order to prevent the infringement of innocent people's liberties. Sometimes it seems that too much effectiveness is sacrificed at the expense of crime victims. Officers, therefore, must understand the system thoroughly and learn how to function within its limitations, still upholding the protection of citizens' rights, honesty, ethics, and honor. Cops have a serious job to do, and great cops take their jobs seriously.

THE CRIMINAL JUSTICE PROCESS

Officers know that a substantial number of crimes are undetected and unreported. This number varies according to the type of crime. During the initial stage of the judicial process the police play a crucial role, having an enormous power of discretion. City ordinances and state statutes offer direction that officers use as guidelines to enforce the laws. All officers are responsible for interpreting the law and applying it to their evaluation of particular situations.

As years pass, rookie cops learn the value and wisdom of discretion; they are the ones who decide whether to begin the criminal justice process. If officers choose not to make an arrest, nothing happens. When probable cause exists, though, strong extenuating circumstances must justify the decision not to arrest. Other judicial employees working in the courts or corrections must follow through on officers' decisions to arrest; they are responsible for responding to the alleged acts.

Once officers investigate the report of a crime, they determine whether there is probable cause that a particular individual committed the offense. If so, an arrest is justified; if not, a report is forwarded that triggers further investigation or documentation.

Following an arrest, the alleged offender must have an initial appearance, or preliminary hearing, if remaining in jail. The initial appearance is a hearing before a judge during which the offender is officially informed of the charges, bail is set, and the offender is officially advised of his or her rights.

If charges against the offender are not dismissed within a specified time period, a preliminary hearing is scheduled, where evidence against the defendant is examined by a judge. Charges may be reduced, or if sufficient evidence exists for further prosecution, the case is forwarded for continued action.

PROSECUTION

Initial appearance, or the preliminary hearing, is a facet of prosecution. The prosecution is sometimes referred to as the prosecutor, district attorney, or

DA. In other areas it is known as the state attorney. Prosecution is vital to the police and justice. Some prosecuting attorneys have excellent reputations, while others are the brunt of criticism.

Prosecutors have virtually total control over the processing of cases. As an example, there seldom is a problem establishing a prima facie (valid, sufficient at first impression) case during preliminary hearings. Defendants often waive their rights to preliminary hearings because prosecutors have so much influence over the pretrial process. Prosecutors decide whether to drop or continue a case. In some cities as many as two-thirds of all cases are reduced. The prosecutor must make and justify such decisions.[4]

One of the challenges of the judicial system is correcting the tendency of the various subunits to not understand or not appreciate the problems of other divisions within the system. For example, the police are unsympathetic toward the problems of the prosecution or the courts and vice versa. A frequent complaint from officers is that prosecutors do not prepare their cases well enough for trial. Many officers can recall instances when it was obvious that a prosecutor was unfamiliar with the facts of a case. One reason for this is that most prosecutors are overworked. This is true in many law enforcement-related offices, but attorneys daily are assigned more cases than they can prosecute adequately. They simply do not have time to prepare or prosecute all cases appropriately. In fact, they often miss depositions or pretrial hearings because they cannot be in more than one place at the same time. For example, after a trial ends prosecutors attend a hearing with a judge and defense attorneys to determine which trial is next. Many times the new trial will begin the same day, leaving little time for preparation.

Every cop needs to understand the significance of prosecutors' work loads, but that is unlikely to ever change. In addition to being overworked, most prosecutors are underpaid and up against the expertise of more experienced defense attorneys. Officers need to appreciate the overwhelming case load of prosecutors and accept it as a fact of life. Typical prosecutors are sincere, dedicated attorneys. They are frustrated with the system, too, but do their best under strained and exasperating conditions.

One of the most productive things officers can do is to understand that the police and prosecution are on the same team and do their best to assist them, not to end their job as soon as the arrest is made. Officers are professionally and ethically obligated to assist in the prosecution of cases; officers must prepare thorough investigations; write concise, thorough reports; prepare for depositions; and deliver professional testimony in order to help convict defendants.

COURTS

The American system of criminal justice was conceived long ago by our forefathers. Their vision was to provide a means of equal protection for all citizens through a due process of law. In their minds it was equally important to provide physical protection to society from individuals who would harm others. Unfortunately, according to citizen opinion studies, most Americans have little confidence in the courts to properly convict and sentence criminals. This apparent ineptness of the judicial system has degraded the American lifestyle.

Some people feel as though they cannot be comfortable and free from harm in their own neighborhoods. Contrary to popular belief, the courts are doing their best. Like other facets of the judicial system, the courts are overwhelmed with cases. There are not enough judges, courtrooms, bailiffs, clerical staff, or public defenders.

Each state has the authority to legislate specific state statutes and develop rules of criminal procedure. This authority is influenced by political climate; therefore, it varies in effectiveness and efficiency. All judicial circuits conform to the Constitution and are influenced by prior-case law. In addition, proceedings vary according to the type of court system. The rules of discovery are substantially different in federal court than they are in state court. Further, procedures differ in misdemeanor and felony systems. As experienced officers know, there is also a substantial difference between adult and juvenile court (a higher concern is placed on the welfare of each juvenile).

Lastly, the preponderance of evidence in civil court is vastly different from that in criminal court. Criminal courts strive to ensure that due process of law is observed. Defendants have the opportunity to confront witnesses; be formally notified of charges against them; and have the right to an impartial jury, legal counsel, and to present evidence on their behalf. Due process is a constitutional right. Though it complicates, slows, and at times is a frustration to swift justice, it is absolutely essential to prevent injustice.

Efforts to protect the innocent have generated more limitations upon prosecutors. Cases must be proven beyond a reasonable doubt. Throughout the years, case law has interpreted the Constitution. In doing so, continued regulation has been placed upon the prosecution. Every officer recognizes names and phrases such as *Miranda, Mapp,* the *exclusionary rule,* and *fruits of the poisonous tree.* Some view such guidelines with disdain or believe these restrictive regulations do nothing but create injustice.

No officer should let frustration with the system cause bad feelings toward prosecutors. Remember, the American system of criminal justice has been developing for more than two centuries. During that time, hundreds of thousands of people have devoted their lives to uphold procedures that are intended to safeguard all citizens.[5] Obviously, it is far from perfect. Many countries do not restrict their police and prosecution to the same extent. On the other hand, many of them send more innocent people to prison. The best that officers can do is commit themselves to know the law, understand the system, and always do their best to live by the law enforcement code.

"It takes a special kind of person to be a police officer. They must remain alert during hours of monotonous patrol yet react quickly when need be; switch instantly from a state of near somnambulism to an adrenaline-filled struggle for survival; learn their patrol area so well they can recognize what's out of the ordinary.

"It takes initiative, effective judgment, and imagination in coping with complex situations—family disturbances, potential suicide, robbery in progress, gory accident, or natural disaster. Officers must be able to size up a situation instantly and react properly, perhaps with a life or death decision.

"Officers need the initiative to perform their functions when their supervisor is miles away, yet they must be able to be part of a strike force team under the direct command of a superior. They must take charge in chaotic situations yet avoid alienating those involved.

"They must be able to identify, single out, and placate an agitator trying to precipitate a riot.

"They must have curiosity tempered with tact, be skillful in questioning a traumatized victim or a suspected perpetrator. They must be brave enough to face an armed criminal, yet tender enough to help a woman deliver a baby. They must maintain a balanced perspective in the face of constant exposure to the worst side of human nature, yet be objective in dealing with special interest groups. "And if that isn't enough, officers must be adept in a variety of psychomotor skills: operating a vehicle in normal and emergency situations; firing weapons accurately in adverse conditions; and strength in applying techniques to defend themselves while apprehending a suspect with a minimum of force.

"Then, when it's all over, they must be able to explain what happened—in writing, to someone who wasn't there, in such a way there's no opportunity for misunderstanding—and to document their actions so they can relate their reasons years later."

Bill Clede
Retired police officer and trainer,
Member/American Society of Law
Enforcement Trainers, Charter member
International Association of Law Enforcement, Firearms Instructors, former Technical
Editor/Law and Order Magazine,
Author/Four Books for Police

CRIME AND JUSTICE

Most law enforcement agencies across the nation complete the Uniform Crime Report (UCR) each year. Statistics for individual communities, states, and the nation are compiled, published, and distributed. UCR statistics document the overall crime rate, individual offense categories, and other areas of concern. When crime rates increase, agencies claim that more officers and equipment are needed, and the statistics are used as justification. When crime rates decrease, chief administrators praise their agencies for their excellent work, and citizens are commended for their community involvement.

Regardless of the number of laws, the crime rate has always been too high. Legislation evolves slowly, reflecting the evolution of society. Legislation responds to a public outcry to control behavior. During the twenties and thirties, robberies, crimes against persons, and prohibition-related offenses were prevalent. After World War II, a new breed of criminals and crime developed. Today assorted frauds, thefts, and computer-related offenses occur. Future overall crime rates will decrease, but be more violent due to drug-related attacks.

Criminologists, sociologists, and those within the criminal justice system generally agree that to even substantially diminish crime for the long term would be an overwhelming task. Nonetheless, there has been a gradual decrease in many rates during recent years, due to the baby boom generation aging and maturing past the crime-prone age group.

Law enforcement cannot stop crime. To a degree, crime prevention units, neighborhood watch associations, antirobbery divisions, and decoy operations make a difference, though it sometimes seems futile. No matter how hard you work or how good a job you do, there will be an endless stream of criminals. Yet, doing one's personal best is one of life's great accomplishments. The lives of great officers have been ruined because they could not accept the fact that they could not change everything that is bad or evil.

To eliminate crime, America must eliminate unemployment, racism, inadequate education, poverty, dysfunctional family life, drug abuse, juvenile delinquency, problems of the mentally ill and elderly, and discrimination and inequality. In other words, we have little possibility of preventing virtually all crime, unless citizens are willing to give up their constitutional rights. Officers, therefore, will continue to struggle with law enforcement and true leaders will never lose sight of what our nation would be like without it.

COURT DECISIONS

The best that officers can do to improve the justice system is to work from within it. This is an honorable goal, yet for all of its problems, the American justice system currently is one of the best judicial systems in the world. Let us examine frustrations associated with some court decisions in our system.

Unfair court decisions frustrate officers because they appear to restrain the police. For example, crucial evidence may be declared inadmissible; confessions may be dismissed; and rapists, murderers, and robbers may go free.

Amendments to the Constitution were originally created for protection of citizens. The Fifth Amendment for example, regards self-incrimination: "No person shall be compelled in any criminal case to be a witness against himself. . . ." The intent of this Amendment was fairness; most people agree that it would be wrong for anyone to be coerced into testifying against himself or herself.

The Fourteenth Amendment states, "No state shall deprive any person of life, liberty or property without due process of law." Once again, no one can argue with this amendment; it makes sense. Decade after decade, the courts struck what they believed was a balance between the right of society to be protected and the rights of the accused. The system emphasized the totality of circumstances surrounding particular situations.

During the 1960s, however, the courts began to substantially adjust their reasoning. As an example, the Miranda decision occurred in 1966. Supreme Court Justice Earl Warren explained the majority reasoning as he wrote "police interrogations were so inherently coercive" that courts should consider to first advise them of their rights.

Other respected legal authorities disagree with the reasoning behind Supreme Court decisions like Miranda. Professor Emeritus Fred Imbau of the Northwestern University School of Law states the impact of the Miranda decision has been devastating to police effectiveness.[6] Others claim that the monumental Supreme Court decisions of the 1960s simply went too far and ignored the rights of victims and favored the accused.

These court decisions created untiring, unprecedented controversy. Public outcry and frustration often rose very high. As years passed, the turmoil gradually began to lessen. The 1970s and 1980s found citizen attitude more accepting, even though numerous victim advocate organizations were created.

During the late 1980s, several court decisions started to move more in favor of the victim. The future political and social climate will determine just how far these decisions will go.

CIVIL SUITS AGAINST LAW ENFORCEMENT

Most cops have an enormous amount of sincerity and would never intentionally violate legal rights or deprive someone of their liberties. America is not plagued with widespread police abuse.

Supreme Court decisions of the 1960s regulated actions of the police strictly. Patrol officers had begun to be the focus of civil suits prior to the 1970s. The threat of civil litigation against law enforcement prevailed in the 1980s and 1990s. The frequency of litigation increased in the 1980s. Case law 42 USC-1983 was passed as a federal law intended to prevent officers from violating defendant's rights and abusing the Constitution "under the color of state law." It expanded civil liability to all levels of a law enforcement agency's hierarchy. Some officers who found themselves the focus of civil suits avoided liability by proving they acted "in good faith." Unfortunately, this often caused further media attacks to fall on their agency. Sometimes superior officers were sued for lack of training and/or supervision.

The ability of law enforcement to withstand the wave of litigation took a serious blow in 1978 when the Supreme Court held that cities violating constitutional rights under local custom, policy, or practice could be held liable under 42 USC-1983. Anyone bringing claims against the police could now challenge a city's customs and policies or traditions. Before long, another ruling held that municipalities could no longer claim the good faith defense to constitutional violations.[7]

From the officer's point of view, the storm of civil litigation against them seems unjust. After all, most cops are on the street doing the best that they can. They are not lawyers who have months to examine a situation and then act accordingly. On the streets, they must often make split-second decisions under stressful conditions.

While it may be unfair to closely examine each move, then sue someone or their agency for every penny possible, it is everyone's right to file suit. Civil litigation against the police is unlikely to change. Always remember that professionalism will defend you. Surprisingly, there have been considerable good results from the increase of civil litigation against the police. Law enforcement is held more accountable for its actions.

The benefit of testifying in criminal court is that every action is examined, which causes officers to reassess their actions. If mistakes were made, they will learn from them. When officers and their departments are sued, likewise it forces everyone to examine themselves, their policies, methods of operation, and the effectiveness of training. A positive effect of litigation against the police is that it has caused law enforcement to become more professional. This has affected sheriffs, chiefs, and county and city administrators.

In February 1989 the Supreme Court ruled that cities can be held liable for some injuries if the concerned police officers' training was inadequate. The *City of Canton vs Harris* ruling expanded the civil liability of local governments. If a failure to train amounts to a "deliberate indifference" to constitutional rights and is responsible for the damages which have occurred, the agency and government can be held accountable.

Good police work can usually prevent and refute false accusations. This requires working the street effectively and being equally conscientious about all actions. Write good reports; never do anything to lose your most prized possession, your integrity; revise departmental procedures when needed; and document effective training. When officers are guilty, they must be responsible for their own actions. When allegations of misconduct and abuse are true, litigation does us all a favor, since such officers should not be on the force.

PLEA BARGAINING

Most criminal convictions are obtained through plea bargaining. Each year thousands of criminals receive reduced sentences, probation, withholding of adjudication, or adjudication of a lesser charge. Often plea bargaining enrages victims, witnesses, or entire communities. Some officers vent their frustration over cases of plea bargaining by becoming enraged or acting out—actions they will regret later. Others, though unhappy, accept the fact.

At times, plea bargaining is a necessary evil or a necessary ingredient within the criminal justice system. In a few isolated instances negotiations between the prosecutor, defense, and judge have resulted in a mockery of justice.

The frustration, anger, and outrage felt about plea bargaining is the result of not understanding or refusing to accept the realities of the American judicial process. Without plea bargaining, many criminals would go unpunished, and others would receive much lighter sentences. There simply are not enough resources to process the case load, and the system and those within it are already overworked, backlogged, and overstressed. Imagine what it would be like if all the cases that ended through plea bargaining were suddenly scheduled for trials.

DEFENSE ATTORNEYS

Few things are more infuriating to dedicated cops than to have their testimonies and cases torn apart and distorted before juries. It is also frustrating for a jury to hear the truth twisted and distorted. Sometimes it seems as though justice is lost in the shuffle as a few defense attorneys appear more concerned with winning and collecting fees from their clients. Some even intentionally withhold evidence or unethically coach witnesses. Fortunately, rules of criminal procedure and courtroom limitations usually prevent defense attorneys from such behavior.

Officers should never lose sight of what truth and justice mean. Few things are more precious and valuable than living with memories of honor and dignity. As with any profession, there will be a few distorted attorneys among the many good ones. If attorneys try to mislead or distort your testimony, stay calm, collected, and factual. Your honesty, professionalism, and sincerity will be apparent, and so will their deceitfulness.

LEGAL GLOSSARY

It is almost impossible to work effectively within the justice system without understanding frequently used terms and common slang phrases. Understanding the vocabulary will help prevent disastrous results for your cases. Know the proper spelling as well to avoid damage to your credibility. Study the following terms to help you build a practical, effective legal vocabulary.

ABROGATE — To repeal, cancel or annul.

ABSCOND — To leave a jurisdiction of the courts to avoid legal process.

ACCESSORY — Any person who has knowledge that an individual has committed a felony or has been charged with a felony and aids that person with the intent that the individual avoid arrest, trial, or conviction. An accessory must have knowledge that the individual has committed, been charged with, or convicted of the felony.

ACCUSED — The defendant in a criminal case.

ACQUITTAL — The formal finding of innocence of an individual charged with an offense.

ADJUDICATE — To make a final determination.

ADVERSARY — An opposing party in an action.

AFFIDAVIT — A sworn written declaration or factual statement given before an individual having authority to administer oaths.

AMEND — To correct an error.

APPEAL — The process of obtaining a review and retrial by a superior court.

ARRAIGNMENT — A legal procedure by which a defendant is informed by the court of the charges against him or her, advised of his or her legal rights, and determines his or her plea.

BENCH WARRANT — A process issued by a court for the arrest of an individual. It is usually issued by a judge when an individual fails to appear for court.

BEYOND A REASONABLE DOUBT — A phrase intended to satisfy the judgment of a jury that a defendant committed the concerned crime. It is typically used as a portion of a statement such as "proved beyond a reasonable doubt."

CAPIAS — A term that refers to several types of documents that require an officer to take the defendant into custody.

CERTIORARI — A document issued by a superior court to a lower court requesting that the record of a case be sent to it for review.

CHANGE OF VENUE — The moving of a trial begun in one jurisdiction to another jurisdiction for trial.

CIRCUMSTANTIAL EVIDENCE — The circumstances surrounding and conditions from which the existence of the main fact may be logically and reasonable inferred.

COERCION — The act of compelling by force or threat.

CONCURRENT — Running together, as in a defendant receiving two sentences which are served concurrently.

CONSECUTIVE — Served one after the other, as in "The defendant received consecutive sentences."

CONVEYANCE — A written paper on which property or title to property is transferred from one individual to another; also a type of vehicle.

CORPUS DELICTI — Basic facts required to prove the commission of a crime.

CORROBORATION — Additional evidence which supports the testimony of a witness.

COUNT — Used to specify more than one part of an indictment or information, each charging a separate offense. Frequently used synonymously with "charge."

CRIMINAL INTENT — The intent to commit a crime.

CURTILAGE — An enclosed space which immediately surrounds a dwelling.

DEPOSITION — A statement obtained through questioning, usually transcribed, signed, and sworn to.

DELIBERATE — Determining the guilt or innocence of a defendant by weighing the evidence by a jury.

DIRECT EVIDENCE — Evidence of proof which tends to show the existence of a fact without considering any other fact.

DUCES TECUM — Latin, meaning bring with you. Generally appearing on a subpoena, requiring someone to appear and bring certain items with them.

DURESS — Forcing someone to do something against their will.

ENTRAPMENT — The act by the police of inducing an individual to commit a crime.

EVIDENCE — The means through which an allegation is established or disproved. Proof which is usually presented at trial for the purpose of establishing belief.

EXCLUSIONARY RULE — A rule which prevents evidence seized illegally from being admitted in a trial.

EX PARTE — Done for or by only one party; on behalf of only one party.

FRUITS OF A CRIME — Material objects obtained through the commission of a crime and which may constitute the subject of a crime.

GROSS NEGLIGENCE — The obvious failure to exercise the extent of care required by particular circumstances.

HABEAS CORPUS — Documents which command someone to produce a detained individual before court.

HEARSAY — Evidence not from the personal knowledge of a witness, but from what has been heard from others.

INDICTMENT — The formal charge of crime issued by a grand jury after hearing evidence surrounding the offense.

INFORMATION — An accusation of a criminal offense against an individual, issued without an indictment.

INFRACTION — A mirror offense; primarily traffic offenses.

IMPEACHMENT — A technique of indicating a witness is not believable. Also, a criminal proceeding intended to remove a public officer from office.

INJUNCTION — A document that prohibits the performing of a particular action.

INQUEST — A legal inquiry by a medical examiner or court into a sudden or unusual death.

JUDGMENT — The official order or sentence of a court in an official proceeding.

LEADING QUESTION — A question which is influential or suggestive when asked to a witness.

MAGISTRATE — A lower court judge.

MENS REA — Guilty knowledge, a wrongful purpose, or criminal intent.

MISDEMEANOR — An offense that is not a felony; a minor offense.

MODUS OPERANDI — The method of operation used by an offender.

PREPONDERANCE OF THE EVIDENCE — Evidence that, when compared to evidence in opposition, produces a more convincing truth.

PRIMA FACIE — Evidence which is sufficient to establish a fact when first considered.

PROBABLE CAUSE — A statement of facts which would lead a reasonable person to believe the accused individual committed the offense.

RES GESTAE — Things done during an entire event; actions or statements immediately after an incident that are considered to be part of the incident.

RESPONDENT — The defendant during an appeal process.

SEQUESTER — To isolate the jury of a legal proceeding; also, to separate witnesses from other witnesses to prevent them from hearing testimony.

SUBPOENA — A document which commands appearance of a witness.

SUBPOENA DUCES TECUM — A document commanding an individual to produce papers for a court proceeding.

UTTERING — To publish or circulate.

VENUE — The location in which a fact is alleged to have occurred; the location in which a court with jurisdiction may hear a legal proceeding.

WRIT — A judicial instrument used by a court to command an action by an individual.

END NOTES

1. Hazel B. Kerper, J.D., *Introduction to the Criminal Justice System,* West Publishing Co., St. Paul, MN, 1972, p. 7–14.
2. Ibid., p. 22–24.
3. John N. Ferdico, J.D., *Criminal Procedure,* West Publishing Co., New York, 1989, p. 4–6.
4. The President's Commission on Law Enforcement and Administration of Justice, *The Challenge of Crime in a Free Society,* U.S. Government Printing Office, Washington, DC, 1967.
5. Ibid.
6. Eugene H. Methvin, "The Case of Common Sense vs. Miranda," *Reader's Digest,* Pleasantville, New York, August 1987, pp. 98–99.
7. Candace McCoy, "Lawsuits Against Police: What Impact Do They Really Have?" *Criminal Law Bulletin,* Warren, Gorham and Lamont, Boston, MA, 1984, pp. 57–58.

"A great cop is a person who understands the nature of policing in a free society. That person should be able to empathise with persons who are victims of crime, be compassionate with those less fortunate, and treat all people, even those accused or convicted of criminal activity, with evenhanded temperance."

J. Dale Mann
Division Director
Georgia Law Enforcement Academy

CHAPTER 5

SURVIVING THE STREET

"To be a great cop is to have the ability to take the job seriously without taking yourself too seriously. A great cop is recognized by his actions and the manner in which he carries himself. It is called ***quiet confidence."***

Peter C. Loomis
Professional Standards Coordinator
Maitland, Florida, Police Department

Cops are good people. Some develop a cynical, tough exterior as a defense against the crime on the street. "The Blue Wall" is comprised of people who risk their welfare for the protection of others, for no other reason than it is just the right thing to do.

If you were to ask someone applying to an agency why he or she wanted to be an officer, ninety-five percent of the time the answer would be "to help others." Their feelings may change as the years pass, because of the frustration from working the street and the injustices of the justice system. It takes a special blend of sincerity and courage to intentionally put yourself in the middle of violent or tragic situations daily. Imagine exposing yourself to brutality, violence, and hatred in order to resolve situations so that others will not be injured. It is an act that requires an equal amount of courage and nobility to survive. Every moment of every day an officer somewhere in the nation is at risk. It may be when a cop is in an alley at 0430 hours, walking up to a vehicle, separating a husband and wife during a disturbance, or arresting someone for assault. Officers do not do this for medals or other recognition; they do it simply because it is their job.

CIRCUMSTANCES OF OFFICER MURDERS

Every several days in America a cop dies. Around eighty officers are murdered each year. About twenty more die in the line of duty by other means. As they take their last breath a little bit of America departs with them. The question is, did they have to die? Is there anything they could have done to survive? In most cases, there were techniques that could have increased their chances of survival. To learn these techniques, every cop should know the answers to the questions when, where, how, why, and by whom.

During the 1970s and 1980s the Federal Bureau of Investigation compiled a revealing ten-year study of American officers and how they are killed in the line of duty. The data was provided using crime-related information compiled and submitted by agencies across the nation.

The study reveals that most officers are killed in the South, and that California, Texas and Florida are also dangerous.

The majority of cops are killed by firearms, and the handgun is by far the most common deadly weapon. The next most frequent weapon used, even though it is a relatively small percentage, is the knife. Bombs, personal weapons, and assorted miscellaneous objects account for the remaining types of weapons.

For more insight concerning the firearms used against officers, the Federal Bureau of Investigation provided a study of officers murdered during 1983. During that year 80 officers were murdered. Of that number, 74 were killed by handguns. Twelve of the handgun deaths were committed using the deceased officer's own gun. The most frequently used handguns by perpetrators were .357 magnum and .38 caliber.

The ten-year Bureau study determined the distance officers were from their perpetrator at the time of death when firearms were used. In the concerned 960 police murders, 494 officers were situated at a distance less than 5 feet away from their assailant. In 192 cases the distance was 6 to 10 feet. In 152 instances it was between 11 and 20 feet. In 66 cases the distance was between 21 and 50 feet. In only 56 cases the distance was more than 50 feet away.

The location of fatal wounds suffered by victim officers was also revealed in the decade-long research. Of the 960 total murders 451 victims suffered front upper torso firearm wounds. Three hundred nine victims suffered from front head wounds, the next most frequent area. All other locations of wounds were relatively infrequent.

Hours of darkness are much more dangerous for officers than daylight. Generally, officer deaths increase the closer it is to midnight. The two-hour period from 10 p.m. until midnight is the most dangerous. Several hours before and after midnight also have significant officer deaths. Few murders occur during the first several hours after sunrise.

Research revealed that there is not a significant difference in officer murders related to the months of the year. Of the total 1,031 deaths, the fewest in any specific month was 65. The most in a month was 102, in December.

The most dangerous day of the week for officers will surprise you. More officers were murdered on Thursday during the ten-year period. The next most dangerous days were Saturday, Friday, and Tuesday, respectively. Significantly fewer deaths occurred on Sunday than on any other day.

Of all the officers killed, most died during arrest situations. The next most dangerous situations were responding to robberies and disturbances.

Contrary to popular belief, young rookies are not the age group murdered most often. Only 116 officers were under twenty-five years old. Three hundred three officers were between twenty-five and thirty years old. Three hundred sixty-nine were between thirty-one and forty. Lastly, 243 officers were over forty. The average years of service at the time of their deaths was eight.[1]

ANTICIPATING DANGER

One of the most beneficial and effective ways officers can safeguard themselves on the street is to develop the ability to anticipate danger. This should not be confused with being scared, timid, or overreacting. A great deal of mental preparation and control are required to ensure that you will not overreact. Most officers resort to the way they have been trained. If you have not been well trained, however, no one can tell what you will do!

To anticipate danger, take full advantage of all your physical senses and be prepared for the unexpected. Like most aspects of police work, survival on the street is a matter of attitude. Stay alert, be aware of your surroundings, and develop effective intuition.

Anticipating danger is an outlook which must be your constant partner. The following examples of techniques and strategies can help make the best survival tactics become a way of life.

- Always think and plan ahead as you respond to a call.
- Know where you are at all times.
- Anticipate the unexpected at all times and in all places.
- Learn how to verbally defuse potentially violent situations.
- Use all your physical senses to their maximum.
- Always have a plan of action in case you are assaulted.
- Be constantly alert to any signs of danger.
- Do not patrol the same way every day.
- Never lose sight of a suspect's hands.
- Always keep your gun hand free and stand so that it is away from all civilians.
- Never become too comfortable on the street.

PSYCHOLOGICAL CHANGES AND REACTIONS IN CRISES

Wherever you happen to be at this moment, whomever you happen to be with, imagine someone bursting into the room and shooting in your direction. Before you would have time to react, your body would undergo enormous physiological changes, so startling that they would substantially alter your ability to react.

Nature provided us with incredible physical abilities. The immediate physiological changes in times of crisis are intended to help us survive. Within a fraction of a moment, blood vessels constrict throughout the body. In times of severe trauma, the constricted vessels assist in reducing blood flow. Suddenly diminished blood vessels cause a chain reaction of additional changes. Because blood carries oxygen to the brain, constricted vessels prevent sufficient oxygen from traveling to it. Therefore, the brain signals the heart to beat harder and stronger to receive more oxygen. All of us can relate to the feeling of a quickened heartbeat during high-anxiety situations.

The brain also signals the lungs to breathe deeper and faster. This forces oxygen into the bloodstream. Some officers recall witnessing people fainting upon receiving death notices. This is caused by the constricted blood vessels suddenly preventing sufficient oxygen to the brain.

Additional changes also occur during crises. A sudden rush of adrenaline provides the body with unusual strength and power. Anyone who has been in an extreme crisis may have experienced "tunnel vision," a phenomenon allowing one to see only what is directly in front. Despite screaming, shouting, or gunfire nearby, the individual probably will not hear anything. Even though the person is running or struggling with every ounce of strength, it can seem like everything is moving in slow motion. Even the eyes dilate to allow better night vision. The body is remarkable, especially when it is kept well-conditioned.

Officers who have been in shoot outs agree they experienced all or many of these sensations. Yet the most important physical change has not been mentioned. The lack of oxygen from less blood flow causes a decreased ability to think clearly, which is so important to survival. Most officers do not even know it occurs, and yet it happens every time a high-stress circumstance arises.

Not being able to think clearly and quickly in a crisis is obviously unwanted, especially in that moment requiring a split-second decision. Yet, your mind may go numb at precisely that moment. Afterwards, as the months pass and the judge, jury, media, and defense attorneys scrutinize every fraction of a second, no one but you will appreciate the difficulty of that brief moment.

If you have ever started to give a speech or respond to questions only to face the sensation that your mind has gone blank, the sensation is due to the stress of being called upon to speak in front of other people. As a result, oxygen flow to your brain is quickly reduced and you cannot think very well. This happens in life-and-death situations. The difference is that instead of not being able to recall something, your mind and body will probably experience total numbing. Senses will be distorted and a feeling of being in slow motion will encompass you.

To combat the physiological phenomena that strips officers of their mental abilities when they are needed the most, understand, appreciate, and expect that it will probably happen. While responding to a call take advantage of every second. Plan ahead; use each moment to plan what you will do when you arrive at the scene. Once you arrive at the scene and the crisis is in motion, you will not be able to think as well. Think about what you are going to do, so that you will not have to live with the consequences.

To better prepare for these physiological phenomena, you must have completed sound training. This will also help you through the rough times; resorting to the way you have been trained is a natural reaction.

GOOD COPS AND BAD COPS

For the purposes of this discussion, the terms "good cops" and "bad cops" have nothing to do with corruption, but with their quality of work. It has everything to do with pride and honor. Good cops care about their jobs and those they are sworn to protect. Bad cops, on the other hand, write sloppy, inadequate reports and never do anything by their own initiative. They complain about everything and just do not care much about anybody or their jobs anymore.

To demonstrate in a practical sense how crucial it is to have a survival state of mind, let us look at how bad cops think when responding to a call. Then we will compare it to the thoughts of a good cop in the same situation. The difference is obvious.

Our example is of a family disturbance call, although the same principle can be applied to any situation. The bad cop hears the dispatcher advise, "Be in route to a family disturbance at 000 Bahama Road." The officer acknowledges the call, drives to the house, handles the call, and later advises the dispatcher that he or she is back in service. The officer has responded to hundreds of such disturbances, and they have become routine. The officer feels that he or she can handle anything that comes along, and there is no reason for concern.

The good cop handles the same call differently. After receiving the dispatch, he or she may ask probing questions. Was the disturbance in progress? Were the parties armed? If so, with what types of weapons? These are the essential facts to know. This officer is planning ahead, thinking survival the moment the call is received. A great street cop takes advantage of the time spent driving to the scene by gathering as much information as possible. It pays off if there is a split-second decision to be made after arrival.

If the dispatcher advises that the disturbance is "hot," the officer would ask for the location to be pinpointed. As an example, knowing that the address is three houses west of a certain cross street on the north side of the road would allow the officer to arrive much faster than if he or she had to drive slowly enough to check for house numbers. In addition, he or she would be able to find the house quickly in the dark, if it were night, even with the vehicle lights off.

Good cops do not just drive to the scene of a crime. They are professionals and every moment is seized and utilized to plan. If they have backup, the officers coordinate the routes and time of their arrival at the scene. The probability of apprehending a fleeing suspect is increased dramatically as a result of arriving at the same time as the backup.

While continuing to drive to the scene of a hot call, good officers would think about everything that may help after arrival. They prepare by trying to remember or discover anything about the people who live at the scene. Are they known to be violent? What type of weapons will they probably have? What does the inside of the house look like and what is the layout of the rooms? What do other officers know about the suspects? Have the suspects had other recent problems?

Once the officer arrives at the scene, he or she parks away from the house to decrease the risk of assault while getting out of the car. He or she parks so the vehicle is between himself or herself and the house. Before getting out, the officer pauses to look at the house, windows, doors, and the surrounding area. Instead of casually walking toward the house, he or she concentrates on the windows and doors, looking for anyone or anything suspicious. If someone appears, the officer focuses on the person's hands. Great cops know that the best survival technique is anticipating danger.

While nearing the house, the officer looks for trees, vehicles, or other objects to use for cover. A great cop follows these procedures because he is a professional. Just as there is a difference in the quality of physicians, mechanics, attorneys or accountants, some officers are extremely good at what they do. This scenario is an example of how they think. Attitude is everything!

When walking to the house, the officer is careful not to pass directly in front of doors or windows and listens for sounds or voices. Also, the officer studies door hinges to determine whether doors will open toward the inside or outside; this helps to avoid the possibility of them being suddenly slammed into him or her. The officer also knows to stand off to one side to avoid being shot. If any doors open, the officer studies them as they begin to open and watches for signs of an attack through the cracks of the doors.

As the example has shown, two officers can respond to the same call in dramatically different ways. The mental aspect of staying alive on the street is very private. No other person except for officers or their partners who work beside each other every day, knows these thoughts. Officers who are conscientious will gradually develop reputations as good cops.

Officers who cannot develop a survival state of mind need help in developing one from a trainer or supervisor. The street is no place to be lackadaisical and careless. If a cop will not maintain a survival attitude for himself or fellow officers, surely it should be done for his or her family. If you have children, think about your kids growing up without you. If you do not have children, then think about your family and friends carrying on their lives without you. Then decide what kind of cop you are going to be.

"Although training and experience are critical factors for achieving excellence, all truly great law enforcement professionals start their careers with the highest level of honesty, courage, integrity, loyalty, reverence for the law, and respect for others. These ideals, combined with an intense desire to serve, create the foundation for all those who bring great honor and respect to the badge."

Bernard C. Parks
Chief of Police
Los Angeles Police Department

DANGEROUS MYTHS

Statistics give us the data we need for effective survival training and awareness, but they cannot replace the wisdom gained through many long years of working the street. All too often misconceptions are born from apathy and indifference. When it is left unchallenged, ignorance can lead to tragic and unnecessary deaths.

The following myths can get you killed! Inexperienced officers or those who are apathetic and careless usually believe them. Be sure to confront officers who feel these statements are true—you may be saving their lives.

- It is not necessary to handcuff children, women, or the elderly.
- All the hype about wearing a vest, especially during the day, is ludicrous. Nothing has ever happened to me, and it never will.
- After you have responded to the same burglary or robbery alarm several times, do not take it seriously.
- Cleaning a firearm is a waste of time. No one ever inspects it, and a dirty firearm is as safe and useful as a clean one.
- Officers who inspect their shotgun at the beginning of their tour of duty are either rookies or insecure.
- Asserting your authority physically and taking names is what good police work is all about.
- Being loyal and dedicated is stupid. You have got to look out for "number one" if you want to get anywhere.
- If the department wants anything from me, it better expect to pay me for it.
- Juveniles today are less dangerous then those of past decades.
- No good street cop is ever afraid of the street. Mental conditioning is nothing more than a way for trainers to make extra money.
- No suspect will ever get away from me in a vehicle pursuit.
- You do not have to worry about getting stabbed if you have your vest on; a knife cannot go through it.
- Real cops do not need a backup.
- Robbery alarms are always false.
- You do not need more than one officer to search a building or make an arrest.
- Checking out a situation before you stop a vehicle or walk up to a suspect means that you cannot handle the job.

MENTAL PREPAREDNESS

In the world of athletics, the best of the best know the value of mental preparedness. Athletes and other professionals train themselves to envision success and achievement. Some individuals refer to it as positive self-imaging, or mental preparedness.

As in the world of athletics, the real winners in life are competent and poised for success. They see themselves accomplishing their goals and subconscious thoughts of failure do not exist. These winners have convinced themselves that they can accomplish anything if they work hard enough and keep the right attitude.

Law enforcement can use the same insight, with initiative. Positive mental process can be applied to any police circumstance. Even before going on duty, pause to contemplate and reaffirm your survival mind-set. Like a world-class athlete, the professional police officer knows how to prepare for the task ahead. Talk to yourself, saying, "I will anticipate danger. I will stay alert. I will not overreact. I will survive."

When responding to a call, mental preparedness is crucial. Even routine calls must be handled as though they are not routine. Expect the unexpected. Think. Have a plan for any situation.

WORKING THE STREET

No rookie can understand how to work the street right after graduation from the academy. Many academies prepare officers, but there are many things that cannot be learned in the classroom. The practical, realistic strategies and techniques that follow usually are learned only through years of experience.

Control With Voice Commands

Voice commands, used effectively, are the "unsung hero" of working the street. Lack of training in the proper use of voice commands can result in tragedy. Most cops have never had the value of voice commands explained or demonstrated, but after learning their benefits, use them faithfully.

Imagine responding to a call of unknown trouble at a convenience store. When you arrive near the rear of the store, you pause to look at the surrounding area then quietly exit your car. You draw your revolver and cautiously walk along the side of the building toward a tree for cover. You are about to have a dispatcher phone the inside of the store when an armed perpetrator suddenly runs around the corner of the building. The perpetrator appears surprised to see you as you assume a combat stance, raise your handgun, and shout as loudly as you can in a mean, ugly tone, "Police! Stop!"

Your expression matches your voice and convinces the perpetrator in a split second that you mean business. If your voice command is an ear-splitting scream, its power will startle the perpetrator. Try this technique in a training class with an instructor. Present the same scenario and role-play (without guns) the part of the officer who shouts. Every person in front of you will jump if your voice is powerful enough.

In the perpetrator scenario, your voice command has accomplished two important things. First, it has intimidated the perpetrator into not testing you. Second, if the perpetrator starts to fire, he or she will be shooting while off balance, and the chances of shooting you are greatly lessened. If an officer in the same situation uses a weak and timid voice command, the perpetrator is more likely to shoot. The officer failed to control the perpetrator, who, in turn, sized up the officer. The likelihood of being tested will usually increase when an officer's voice and body language do not convey that he or she is someone who is taking charge of the situation. It is obvious to the perpetrator that he or she can get something through violence or threats. After all, he or she has probably lived with or used intimidation extensively.

The use of aggressive voice commands also alerts everyone in the immediate area of danger. Innocent bystanders will quickly scatter as you are verbally stunning the perpetrator. Additionally, the use of voice commands does not interfere with drawing and/or firing your firearm.

"POLICE! STOP!" is a command recommended by some authorities because it is brief, direct, and instills quick reaction. Some officers, especially rookies, like to use a long string of words mixed with profanity. This may be suitable for movies or television programs, but in real life it is unprofessional, hurts your credibility in court, wastes time, and requires preplanning.

Most situations do not require aggressive voice commands. In fact, the majority of police encounters can be controlled or defused with speech that de-escalates potentially emotional situations. In situations such as traffic stops, family disturbances, or neighborhood quarrels, words that control rather than incite flared tempers are best. Keep in mind that you do not need to take a lot of verbal abuse. Rather, the art of "verbal communication" is a useful skill not easily acquired, but worth the effort.

Telegraphing

The word *telegraph* means to make known by signs or gestures, to communicate or make aware. In law enforcement, to telegraph means doing things or making movements that let a suspect know where you are, such as on the other side of a wall or about to come around a corner. It is usually done thoughtlessly by putting the barrel of a handgun beyond your position of cover and concealment or looking around an object with a brimmed cap or hat protruding several inches. For cops, not telegraphing is a vital survival tactic that can save their lives.

Street cops will inevitably face situations where the potential for sudden attack literally lurks around a corner. They may or may not have reason to believe that a suspect is concealed or waiting to attack. Proper training will allow them to respond without overreacting and reduce their risk.

Think of what a scared perpetrator with a 9mm will do when spotting the barrel of your gun or brim of a cap coming around a corner. There is a good chance that he or she will try to shoot—at your head. When you telegraph your presence, the suspect has just enough time to sight you and shoot a weapon. It does not matter if he is in a hallway, behind a tree, or under a truck—the probability of your death has increased dramatically. It is wise to practice stillness with an experienced, knowledgeable survival trainer in order to appreciate the dangers of telegraphing (but do so without firearms).

Building Searches

Building searches usually take place at night. Although the events that led to conducting the search may be typical, the search should never be considered routine. Open doors in buildings may mean they were left innocently unlocked, or just the opposite may be true, and an armed perpetrator may be hiding inside. Officers assigned to search the premises of buildings should do so as if a perpetrator is inside. They should anticipate danger.

Seasoned officers sometimes do not take building searches seriously. All too often they do little more than walk through the rooms and hallways. To make matters worse, rookie officers can be influenced by lackadaisical or arrogant veterans, even though the academy and FTO program taught them properly.

A serious mistake takes place even before the door is opened—searching without backup. Some officers, even though assistance is available, insist on searching alone. Unless it is impossible to get any assistance, wait for backup. Searching a building alone shows an indifference to staying alive.

While still outside any door, stand off to the side. The officer nearest the door knob should open the door. With guns drawn, quickly move inside to each side of the doorway. Pause to listen and let your eyes adjust to the change in light. If you are wearing a cap or hat, turn the bill to the back of your head, if agency policies permit.

When entering a building, looking around a corner, or going into a room, do not look, but peek. Glance only long enough to comprehend what you see, not long enough to give someone a chance to shoot you in the head. It takes a fraction of a second to comprehend what is on the other side of a doorway. If you are not sure what is there, peek again at a different height.

Your first peek should never be at eye level. A perpetrator preparing to shoot you usually aims around five feet high because that is the height at which she or he expects your head to appear. It takes little effort to squat and peek around a corner at three feet. If you need a second glance, stand up so that you are at a different height. If you see an armed subject, get behind cover and request assistance—do not be foolhardy.

Use Light Effectively for Safety

Light can help you stay alive, and it can be the cause of your death. Understanding what helps you stay safe in terms of light is nothing more than common sense. We just discussed pausing inside a darkened room to allow for your eyes to adjust. This is a simple, effective, and logical survival technique, and there are many more.

It is a common dangerous practice to silhouette yourself in a darkened area. You should use quick flashes of a flashlight rather than leave it on continuously. Hold a flashlight far enough in front of you so that "backlight" does not illuminate you or your uniform. Do not carry your flashlight in your gun hand.

The proper use of flashlights is obviously crucial to street survival. As an example, since you will have another officer with you during a building search, you should be positioned on opposite sides of a doorway before entering a darkened room. Crouch down, turn the flashlight on, then roll it across the floor to your partner. The light will illuminate half of the room as it rolls to your fellow officer. When your partner rolls it back to you, it will light the other part of the room. Avoid rolling the flashlight into the middle of the room because it may shine back on you.

When peeking around a corner in darkness, squat and hold your flashlight high above your head. Synchronize flashing the light at the same time you peek around the corner.

Use a flashlight that has a push-button switch in addition to the permanent light switch. A push button will reduce the chance of keeping the light on too long. It also prevents the possibility of illuminating yourself if it is dropped during an assault. Remember, your gun should be the only thing carried in your gun hand.

Light can be a devastating enemy if it makes you visible. On the other hand, it can be an asset that works for you. You must decide whether it will be an ally or enemy. For example, a nighttime traffic stop is much safer when you aim the headlights and spotlight inside and slightly to the left of the vehicle. You will be able to see the occupants better, and if any of them should turn around, they would be blinded by the light. In addition, it is wise to order the driver, via your loudspeaker, to turn on the inside dome light.

Position Yourself Safely

Officers walk up to potential suspects all the time, which can be risky. As you walk up to someone, use all of your senses. Never lose sight of the suspect's hands, and watch the suspect's eyes, which can tell you much more than just where he or she is looking. Sense the suspect's overall demeanor; the value of an officer's extra senses, or intuition, should never be underestimated. Is the suspect overly nervous or about to run? Do you sense that there is something wrong? Although intuition is difficult to explain in court or to newspaper reporters, it can keep you alive—do not ignore it.

When approaching a subject, walk to his or her right front. Stand at a 45-degree angle from the front of the suspect's right hand to allow you to see his or her front and right sides. In doing so, you are making probability work for you. Most people are right-handed, and if a suspect attempts to assault you, the chances are that the attack will come from the right hand. Position yourself so that you can see the right side of his or her body, but still be able to observe the left hand. Look for signs of objects that may be weapons. If you realize he is left-handed, simply move to his left side.

Make sure you stand at a distance beyond the suspect's leg reach as well as arm's reach. It takes only a moment for the suspect to reach you if attacking. Never be lulled into a false sense of security because you are beyond arm's reach.

The first thing an officer says to a suspect is usually, "Can I see some kind of identification?" Many officers are shot at this point. Make the subject lean far forward when handing you identification; do not lean forward yourself. This will cause the suspect to be off balance, rather than you being in an awkward position and vulnerable to attack. If you are still suspicious, have the suspect place the identification on the ground and then back away before you pick it up. Do not overreact, but be cautious. Officers have mistakenly shot individuals because they appeared to be reaching for a handgun.

Rookies are taught to stand with their gun side away from a person standing next to them. This decreases accessibility to your sidearm.

To avoid being kicked in the groin, stand at an angle to the suspect. This also limits vulnerability to other vital areas of your body. The sides of your thighs, arms and shoulders can take a great deal more punishment than your abdomen, chest, and throat. Your agility, balance, and ability to respond to an assault are also vastly improved while standing at an angle.

SHOOT OUTS

NCIC, the National Crime Information Center, sends a notice to all agencies when an officer is killed in the line of duty. The notice contains a summary of the circumstances surrounding the officer's death and many departments read it during roll call. As a result, most street cops learn the circumstances surrounding the death of every officer in America.

Only recently has the data concerning these deaths been used as the foundation for effective survival training. Prior to the 1980s, many misconceptions existed about how cops die. The history of firearms training is one of the saddest realities of our profession and directly reflected these misconceptions. Here are some of the most devastating facts.

- Most officers are shot less than four yards away from their assailant.
- The average time of a shoot out is 2.5–2.7 seconds, very unlike what the movies portray.
- The number of rounds fired is usually fewer than three.
- Approximately forty percent of the time when officers are in lethal shoot outs, they will be facing more than one perpetrator.
- Two out of three times a police shoot out will be during hours of darkness, even though most firearms training is still conducted in daylight.

Officers usually have little or no warning when they are being shot. Studies reveal that the mental and physical process for the assailant to fire a firearm takes approximately one-half of a second. Officers generally take slightly over one second to draw, point, and shoot. Cops who believe they will have the time to draw, aim, and shoot are more likely to die in a shoot out; adding the time for aiming brings the total average time to almost three seconds.[2]

Target Center Mass

Most shoot outs are over within a few seconds. The difference between life and death is measured in fractions of a second, and there is no time to aim. You must draw and point as one action and promptly fire as the second action. Your life is at stake and you cannot be concerned with aiming for a leg or an arm rather than killing the suspect. You simply have no time to select a particular target area, and you cannot afford to hesitate while thinking about it.

The term *center mass* refers to the midtorso area of the body. Point your weapon at the perpetrator's midchest region, because firing into the center mass provides you with maximum stopping power. As projectiles enter the perpetrator's chest, the energy surrounding and immediately behind them should stop the suspect by damaging the vital organs. You are not trying to kill the perpetrator, but to stop him or her, and the center mass strategy accomplishes the purpose most effectively.

Shoot to Stop

Advising officers to fire into a suspect's chest while stating that their purpose is not to kill may seem contradictory. No officer should intend to kill a perpetrator before, during, or after firing. Realistically, the probability of killing a suspect after placing two rounds into the chest is high, but killing should never be the intent.

To fully appreciate this point, think about the following scenario. A uniformed officer has been sued concerning an incident where a homeowner was shot and killed. During cross-examination, the attorney for the plaintiff abruptly turns and stares at the officer. She asks, "Officer Jones, you just testified that the incident was a case of mistaken identity. You thought Mr. Johnson, the homeowner, was a burglar. You felt the flashlight he had in his hand was a gun. When he raised the flashlight, you thought he was going to shoot. You testified that at the moment you fired your intent was to kill him to save your own life. Isn't that correct? You wanted him to die, didn't you? Isn't that right, officer? Is that what you testified to? You testified your intent was to kill him, correct? Why? I am sure not only the jury, but his children and wife in the back of the courtroom would really like to hear your answer. So . . . what is it?"

As the scenario points out, civil liability is a stark reality to contemporary police work. The plaintiff's attorney was able to make the officer appear to be a monster during cross-examination. She was able to because the officer had been mentally conditioned to "shoot to kill" when firing is necessary. This attitude is merely a frame of mind.

When you must fire, do it speedily and accurately. Shoot into center mass, but do it with the intent of only stopping the person. Testifying that you only intended to stop the person because his or her actions forced you to fire could mean a great deal to you, your family, and even the perpetrator's family.

Every cop in America, regardless of past experiences, undergoes a split-second decision-making process when confronted with a potential shoot out.

The decision made during that fraction of a second will determine who lives or dies. Regardless of training, experiences, or attitudes, the probability of momentarily pausing because you are concerned with killing another person is greater with the shoot-to-kill attitude. In a shoot out, you do not have time to pause; actions must be quick and decisive. When officers develop a shoot-to-stop viewpoint instead of shoot-to-kill, the possibility of a lethal pause is substantially reduced, and so is the probability of dying.

Training Methods

In a shoot out, officers fire when, how, and the number of times they have been trained, whether through hands-on, interactive training, simulations, or video-driven laser simulation training systems. Unfortunately, some trainers still do not understand the need to anchor desired responses into long-term memory through interactive firearms training. Sometimes the problem is not the trainer; rather, it is an administrator who will not acquire the firearms training program for the trainer to do a sufficient job.

Some police survival trainers believe that firing two quick shots at a time is the best way to respond in a shoot out. However, officers involved in shoot outs must conserve ammunition. Approximately forty percent of the time officers are in shoot outs, they face two or more armed perpetrators. If they are not adequately trained, they may waste ammunition and perhaps die trying to reload.

LEARN FROM NONSURVIVORS

Until the last two decades, few areas in American policing have been as lacking as police survival training. Training effectiveness improved dramatically during the 1980s, and law enforcement should be deeply grateful to those whose initiative and dedication made such tremendous improvement during the 1990s.

The way officers are trained is the way they will react in crises. The following facts are things we have learned from nonsurvivors in law enforcement. The tips provided with them will help you survive the street today.

FACT: Most police shoot outs occur within 12 feet of the perpetrator.
While we were being shot and killed at a distance of a few feet, we were trained for over a century only at distances of 21, 45, 75 and 150 feet.

FACT: You are always outwitted when facing a perpetrator if you have doubts about your abilities.
Strengthen your skills with practice and intense training.

FACT: Always handcuff before you search!
The reverse was taught for decades.

FACT: Approximately forty percent of the time that a shoot out occurs, the officer is facing more than one armed person.
Until the last decade, most agencies conducted firearms training using a single, stationary silhouette target.

FACT: No matter what you have in your hands, when a perpetrator draws a gun, drop what is in your hands and draw your own firearm.
Do not take the time to put away what you are holding; just drop it on the ground!

FACT: Two out of three shoot outs occur during darkness or reduced lighting.
Firearms training for most agencies has been deficient in this area. The proper training will include night firing two out of three firing range practices. Logistics will make this difficult, but when you consider what is at stake, the obstacles must be overcome.

FACT: Never forget that anyone—no matter what size, height, age, or sex—has the ability to take your firearm away if you do not anticipate danger.
Take every call you respond to seriously.

FACT: Sixty-five percent of all the officers murdered with their own firearm in 1997 had one handcuff on the suspect when they were attacked.
Handcuff a suspect immediately.

FACT: In the past officers were taught to fire six shots in succession, to unload the empty shells into their hand, and to turn around and throw the empty casings in a bucket.
Training today simulates firing more realistically concerning when, how, and the number of rounds to fire. Drop the empty casings to the ground!

FACT: Many street and corrections officers have been attacked because prisoners have learned how to protect their faces from pepper spray.
Use caution when approaching people you have sprayed; do not assume it has incapacitated them.

FACT: Making arrests of two people with no backup leaves one unsecured perpetrator.
Carry a hidden plastic cuff or an extra pair of handcuffs at all times.

FACT: Helmets can be a brutal head-butt weapon.
Always ask bikers and motorcyclists to remove their helmets and put them out of reach.

Training officers have done a great job of overcoming realistic in-service survival training. Now that we have reached a new era in law enforcement, contemporary trainers are doing a great job of overcoming ineffective survival training. Today's trainers know that developing realistic in-service training is the key to keeping officers alive. Great survival training includes:

- Assessing your agency's survival training needs
- Writing a good lesson plan
- Keeping instruction standardized
- Using performance goals and objectives
- Instilling realism
- Conducting pretests and posttests
- Having instructors who are passionate about teaching
- Doing remedial training whenever it is needed
- Keeping accurate documentation

NEVER GIVE UP!

Good training helps to keep cops alive. Training, though, cannot give you the will to survive.

The street can be ugly and unforgiving, where officers see the worst of the worst. Many street officers have seen their own blood in the dirt. However, just because you are wounded does not mean you are going to die. Some officers have died only because they convinced themselves they were going to die once they were wounded, even though their wounds were superficial. If you become excited and hysterical, this can cause you to go into shock. Do not lose your will to survive or you may die needlessly.

The bad guys do not follow rules during a street fight. You will need defensive tactics, good physical conditioning, guts, and determination to survive. Keep thinking that your family is not going to lose you. If you realize you have just been injured, do not give up—fight back! Go home to your family by being mentally prepared to survive.

END NOTES

1. Federal Bureau of Investigation, *Law Enforcement Officers Feloniously Killed,* 1974–1983, Uniform Crime Reports, U.S. Department of Justice, Washington, D.C., 1984.
2. Ronald J. Adams, Thomas M. McTernan, and Charles Remsberg, *Street Survival Tactics for Armed Encounters,* Caliber Press, Caliber Press, Inc., Northbrook, IL, 1980, pp. 24–39.

CHAPTER 6

EMOTIONALLY SURVIVING

"A great cop is a person who can be trusted to do the right thing even when no one is looking. It is a person who works hard and treats others as he or she wants to be treated."

Lee D. Donohue, Sr.
Chief of Police
Honolulu Police Department

Many people believe that being a cop is more dangerous emotionally than physically, and they may be right. Criminologists, police psychologists, law enforcement authors, police administrators, and seasoned officers have voiced their opinion time and time again that stress has a variety of severe detrimental effects. This view is different than decades ago, when little or no concern was given to officers' emotional well-being.

Gradually law enforcement has come to realize the significance of the consequences of stress. Without deep appreciation, no effective remedies will occur. Law enforcement in general has done a poor job of taking care of its own, and psychological assistance varies greatly from agency to agency. Regardless of how much support an agency provides, nothing should prevent officers from learning how to successfully cope with the pressures of life as cops. They can live happier, healthier lives if they have the tools to help them cope with the emotions and stress.

LIFE PATTERNS

Gail Sheehy, a highly respected authority on adult development, shares her insight on the stages of adult development in the book *Passages.* Patterns exist in our lives, which, once understood, are a tremendous benefit. The patterns are comprised of predictable steps associated with age groups. During each time period are crises, which adults should expect to experience. Once understood, each crisis can be dealt with effectively. Individuals can then concentrate on reaching their greatest potential, instead of allowing emotions to create chaos and misery.

As adults pass through each of life's stages, they gradually experience subtle changes in their views: what is valuable, how much time should be devoted to work or entertainment, relationships with others, health and safety, religious beliefs, and financial concerns. As we grow older and mature, these stages have been shaped by our life's experiences.

The Twenties

As opposed to the turmoil of adolescence, this period of life is more stable and may seem somewhat longer. It is a time of serious preparation for and beginning of a chosen career. To shape the vision of dreams and accomplishments yet to come gives tremendous personal satisfaction. Frequently, a lifestyle chosen during our twenties will follow us throughout life. The decisions, habits, and patterns established in this stage set in motion a pattern that is not easy to change.

One of the difficulties with this period is the belief that the choices we make are irrevocable. No one ever promised that life would be fair or things would always go our way. When tragedy strikes or serious mistakes are made, optimism and tenacity are crucial. The ability to view setbacks as only temporary and continue on does make a huge difference.

Intimate, long-term relationships are of major importance. To a large extent the values and beliefs developed as a child and modeled by parents will significantly affect the success of a marriage. The desire to "settle down" with a family is mixed with the need to establish one's self firmly within a career.

The Thirties

For many people their thirties are marked with impatience and a feeling of being too restricted. The decisions, choices, and preferences of the twenties often do not seem quite right. The return to relatively minor turbulence forces new commitments. Outgrowing careers or relationships with spouses sometimes leads to devastating problems. The need to improve professionally and personally continues. If single, the thought of living alone for the rest of their lives can create a great deal of anxiety.

To have a meaningful, financially rewarding career becomes very important as families grow and thoughts turn to the realities of retirement. Buying a home, raising children, and settling down becomes the norm.

The mid- to late thirties usually pose new challenges. Often the realization that life is now half over, reduced physical abilities, the loss of a youthful appearance, increased medical problems, and lost dreams are difficult to accept and deal with. Many people compensate for these feelings by reevaluating their immediate and long-range goals. Women usually become aware of this crossroad before men. A mixture of anxiety, exhilaration, and new assertiveness frequently result. For men, nearing forty means time is running out. Self-criticism, placing blame, and a general feeling of exasperation with their jobs frequently lead to difficulty within the family. Everything considered, however, this period of life can be very rewarding.

The Forties

Entering the forties is highlighted by the satisfaction of many accomplishments and continued feelings of restlessness and frustration. More years of concentrated effort on a career are ahead. Many men start developing second careers; others begin to shift their focus away from a career to personal or marriage-related matters.

People who understand the need for personal growth and change cope accordingly, and stability is regained. However, people refusing to take life in stride find failure. The midlife crisis is harder for individuals who withdraw within themselves. Support and assistance from close friends and relatives is crucial; alienating them can be devastating, as the value of family and friends increases with age.

The Fifties

For some people a personal crisis characterized by a feeling of stagnation may reoccur around fifty. If so, the remedy is the same as if it occurred during the forties—a revitalization and realignment of personal beliefs and goals is in order, and counseling is helpful.

Once the renewed purpose has occurred, individuals find a great deal more satisfaction and happiness. The fifties and beyond can be a very rewarding time of life. The key is being able to like yourself and accept life for what it is. A certain mellowing takes place in the fifties. Close friends and family become extremely important, and reminiscing on special memories prompts initiative to seek out the best the future has to offer.[1]

UNDERSTAND YOUR CAREER CHOICE

Considerable research has been conducted in criminal justice academic circles concerning the factors that influence individuals who choose a law enforcement career. It helps to understand why people become officers in order to appreciate the emotional trauma they often experience during a law enforcement career. Self-knowledge will help you get the most out of life. If you want to be a cop but do not know how to survive emotionally, the chances are great that you are in for more than your share of misery and a disappointing career.

Women and men enter law enforcement for similar reasons. Two primary reasons are the desire to help others and job security. Also frequently listed are doing something about crime, excitement, prestige, and an interesting life. Documented research shows that most people become cops because they have a sincere desire to help others. This sincerity toward a worthy purpose makes it difficult for officers when they realize that they cannot rid the world of all of its evil.

The belief that officers want to have power and authority over others is not substantiated by sound research. Studies indicate that such factors do not motivate most people who choose a law enforcement career. Nonetheless, experienced officers have seen more than one rookie with "John Wayne Syndrome."[2] This is a phenomenon whereby officers believe themselves to be indestructible. This feeling that nothing could happen to them often prompts foolhardy actions that may cause injury to themselves or others.

STRESS

The last decade has witnessed endless articles, books, and television specials about stress. The most commonly held notions are that stress is negative. However, stress can be either positive or negative. In general, it is defined as "the body's unspecific response to any demand placed upon it."[3]

Stress is derived from change, both positive and negative. Change is an inevitable aspect of life, so stress is just as certain. Positive stress triggers include events like holidays, promotions, or reunions. Negative stress triggers include divorce, death of a spouse, or losing a job.

Massachusetts Institute of Technology psychologist Dean Ormish notes that the greatest cause of stress is the perceived mismatch between where and who people are and where and who they think they should be. In other words, people's expectations of themselves can produce devastating results. Some people were raised in households where near-perfection was expected of the children. These people as adults have difficulty coping with daily pressures. Hectic days, minor problems, and life's little irritations may be too much to handle unless they, like everyone, learn to take things in stride. Even substantial disappointments have to be viewed as only temporary setbacks in a long line of accomplishments.[4]

Police behavior is a frequent topic of discussion these days. Their behavior is affected by numerous factors, such as negative and positive stress, society's expectations, personal aspects, management, ethics, and more. Cops must exercise care in all these areas in order to avoid tainting their behavior. Kevin M. Gilmartin and John J. Harris compiled an article (see Figure 6.1) that helps us understand how the transition from a good, honest cop to a compromised officer can occur.

FIGURE 6.1

"LAW ENFORCEMENT ETHICS . . . THE CONTINUUM OF COMPROMISE"

Kevin M. Gilmartin, Ph.D
John (Jack) J. Harris, M.Ed.

During the past few years, law enforcement behavior has been the subject of increased scrutiny across the country. Rodney King, Ruby Ridge, Waco, evidence planting in Philadelphia, Mark Furhman's testimony, Operation Big Spender, and the chase and apprehension of the illegal aliens in Southern California are just some of the incidents that have captured the nation's attention. With each new headline, mistrust of law enforcement increases; police/community relations suffer; and the reputations of good, hardworking and ethical law enforcement professionals and their organizations are tainted. Even the most avid supporters of law enforcement wonder what is happening and are asking, "Can the police be trusted to police themselves?" While high-profile cases capture the nation's attention, law enforcement agencies across the country spend an increasing amount of time investigating, disciplining and prosecuting officers for unethical or criminal behaviors that never make it to the front pages.

Is the concern over inappropriate police behavior just sensationalized media coverage; have a relatively few number of incidents been used to taint an entire profession or is this a real problem that needs close attention and immediate action? Unfortunately, the incidents that have made the headlines have tainted the reputation and called into question the behavior of the entire law enforcement community. These highly publicized incidents do not, however, address the more subtle ethical dilemmas that law enforcement agencies and their communities have to face each day. Law enforcement agencies across the country face issues of integrity that have to be examined and changed.

Concerns about police integrity, corruption and unethical behavior are not new. The Mollen, Knapp, Christopher and other commissions have examined issues in their respective jurisdictions. Suggestions and recommendations that have come from these and other commissions and investigations, although insightful or accurate, have not resulted in significant acceptance or change within law enforcement organizations nationwide. While departments may have increased the number of required ethics-related classes, the training itself has not changed dramatically. The information is still not internalized nor appreciated at the street level, or throughout the organization for that matter. For many officers, ethics training is seen as nothing more than a politically driven, knee-jerk reaction to the media attention that surrounds high-profile cases. Significant changes in the way law enforcement ethics is conceptualized, taught, and integrated throughout an organization are needed. Without these changes, it is doubtful any information will be accepted and internalized into the day-to-day lives of police officers and police organizations or result in meaningful change.

Ethics is typically taught during the basic academy or at in-service training after embarrassing situations erupt. It is often seen by instructors and students alike as a class that "has to be taught," but one which

continued

nobody really wants to talk about. While the training is necessary, its importance becomes diluted or rendered ineffective by the manner in which it is presented and/or the socialization process that occurs during a police officer's first year or two on the job.

Police corruption is often seen as a distant problem peculiar to "big city cops" or "other departments." Denial and refusal to accept the potential for ethical compromise and corruption at "our department" prevent administrators and officers from developing an in-depth understanding and appreciation of the issues. Without a clear understanding, adequate information and practical strategies, officers who are exposed to a risk-filled environment are more likely to engage in inappropriate behaviors that can destroy their organizations. The transformation from idealistic, highly ethical officer into a self-serving individual who believes "If we don't look out for ourselves, who will?" is a subtle process that usually occurs before the officer knows what has happened. For ethics training to be effective, officers have to see the information as relevant and credible. This approach, even when the information is interesting and enlightening, is rarely internalized by the officers nor incorporated into their day-to-day activities.

The Continuum of Compromise

In this article, the authors explain the "continuum of compromise" (Gilmartin & Harris, 1995). It is a framework for understanding and teaching how the transition from "honest cop" to "compromised officer" can occur. Law enforcement agencies can help prepare their officers for the ethical challenges they will face during their careers. However, that will require changing the way this topic is approached by the organization and teaching and integrating the information throughout the organization.

Officers live and work in a constantly changing and dynamically social context in which they are exposed to a myriad of ethical conflicts. When either unprepared or unaware, officers are more likely to "go with the flow" than they would be if they were adequately prepared to face potentially ethical risks. Every day officers practice mental preparation as it relates to tactical situations. Officers who are mentally prepared to face a lethal encounter are more likely to be successful than officers who are tactically proficient but mentally unprepared. Just like lethal encounters, ethical dilemmas occur at the most inopportune times, frequently without warning and with little time to stop and think about the situation. When inadequately prepared, even the most honest, above-reproach officers can make inappropriate split-second ethical decisions . . . decisions that can result in life-changing consequences. If officers are going to survive ethical dilemmas they need to be as mentally prepared as they would be for tactical encounters.

While police work is seductive and exhilarating, it can also lead officers down the path of ethical compromise. The "continuum of compromise" outlines the path of ethical compromise and can be used to help officers understand and mentally prepare for the ethical dilemmas they will face. Understanding the issues and being mentally prepared will help officers assume responsibility for and make more appropriate decisions. Compromising behavior has to be seen as something that can potentially affect all law enforcement officers . . . not just those in "corruption rich" environments. Officers who view compromise or corruption as an "all or none" phenomenon will not see compromise as an unlikely event, training will be viewed as a waste of time and officers will not become mentally prepared. Understanding the continuum of compromise will allow officers to recognize the risks, assess their own potential for compromise and develop an effective strategy to ensure ethical integrity. When teaching ethics the goal must be to develop an understanding of the progression towards compromise and the development of self-monitoring strategies to prevent becoming embroiled in compromising events.

The Continuum of Compromise®

A Perceived Sense of Victimization Can Lead to the
Rationalization and Justification of:
Acts of Omission
Acts of Commission–Administrative
Acts of Commission–Criminal
Entitlement Versus Accountability
Loyalty Versus Integrity

A Perceived Sense of Victimization

Officers frequently develop a perceived sense of victimization over time. Officers typically begin their careers as enthusiastic, highly motivated people. However, when these young officers over-invest in and over-identify with their professional role they will develop a sense of singular-identity based on their job and an increased sense of victimization. At greatest risk are officers whose jobs literally become their lives. For them, "I am a cop" is not just a cliché but rather a way of life. Over-identification and over-investment causes people to link their sense of self to their police role, a role they do not control. While this builds camaraderie, it can also cause officers to eventually hate and resent the job they once loved.

While officers have absolute control over their own integrity and professionalism, the rest of their police role is controlled by someone else. Department rules, procedures, policies, equipment, budget allocations, assignments, dress codes, and many other day-to-day and long-term activities are controlled by the chief, commanders, supervisors, prosecuting attorneys, the criminal justice system, laws, the courts, politicians, etc. Officers who over-identify with the job soon experience a loss of control over other aspects of their lives. Professional over-investment, coupled with a loss of personal control puts officers at serious risk, a risk that in some ways is more dangerous than the physical risk they face on the street. "It doesn't matter how guilty you are, but how slick your lawyer is," can become the officers' cynical yet reality based perception of the legal system. These realities combine with over-investment to develop an "us versus them" perception in terms of how officers see the world.

The physical risks that officers are exposed to each day require them to see the world as potentially lethal. To survive, they have to develop

continued

a "hypervigilant" (Gilmartin, 1984) mind-set. Hypervigilance coupled with over-investment leads officers to believe the only person you can really trust is another cop . . . a "real cop" that is, not some "pencil-neck in the administration." While officers first become alienated from the public, they can soon distance themselves from the criminal justice system and finally from their own department administration. "I can handle the morons on the street; I just can't handle the morons in the administration," is often heard among officers. It is ironic how quickly idealism and trust in the administration can change . . . oftentimes even before the first set of uniforms wears out. As a sense of perceived victimization intensifies, officers become more distrusting and resentful of anyone who controls their job role. At this point, without any conscious awareness and certainly without any unethical intent, unsuspecting officers can begin a journey down the continuum of compromise.

As the over-invested officer detaches from nonwork-related interests or activities, a perceived sense of victimization will increase. Peer groups, friends, co-workers and potentially their entire frame of reference of life begins to change. By itself, feeling like a victim is by no means equivalent to being ethically compromised. However, feeling like a victim (whether real or imagined) is the first stop on the continuum of compromise.

Acts of Omission

When officers (or anyone for that matter) feel victimized, in their own mind they can rationalize and justify behaviors they may not normally engage in. "Acts of Omission" occur when officers rationalize and justify not doing things they are responsible for doing. At this point, officers can feel quite justified in not doing things that, from their own perspective, appear to "even the score." "If they (whomever it may be) don't care about us, why should we care about them." Acts of omission can include selective nonproductivity (ignoring traffic violations or certain criminal violations, etc.), "not seeing" or avoiding on-sight activity, superficial investigations, omitting paperwork, lack of follow up, doing enough to just "get by" and many other activities which officers can easily omit. "You will never get in trouble for the stop you don't make!" typifies the mind-set of officers during this stage.

This results in decreased productivity and produces passive resistance to organizational mandates. "Acts of Omission" rarely face critical scrutiny from peers who themselves are frequently experiencing the same sense of victimization and socialization process. **Peer** acceptance and loyalty become more important than following some arbitrary set of professional principles. The perceived sense of being victimized can allow officers to rationalize and justify other acts of omission such as not reporting another officer's inappropriate behavior (sometimes regardless of how extreme or criminal the behavior may be).

Acts of Commission–Administrative

Once officers routinely omit job responsibilities, the journey to the next step is not a difficult one to make . . . "Acts of Commission–Administrative." Instead of just omitting duties and responsibilities, officers commit administrative violations. Breaking small rules, that seem inconsequential or which stand in the way of "real police work" is the first step. This can set the stage for continued progression down the continuum. Acts of administrative commission are seen in many ways . . . carrying unauthorized equipment and/or weapons, engaging in prohibited pursuits and other activities, drinking on duty, romantic interludes at work, not reporting accidents and firing warning shots are just a few examples. Department sanctions are typically the only risk that officers will face at this point. For most officers this is the extent of their personal journey down the continuum of compromise. Acts of omission and acts of administrative commission are significant in terms of professional accountability and personal integrity. When discovered, they can erode community trust and damage police/community relations. However, they rarely place officers at risk for criminal prosecution. The initially honest and highly motivated officers can now rationalize their behavior along the lines of "I'm not a naïve rookie out trying to change the world . . . I know what it's really like on the streets and we (the police) have to look out for each other because no one else will."

Acts of Commission–Criminal

Unsuspecting officers can unwittingly travel to the next and final stage of the continuum . . . "Acts of Commission–Criminal." In the final stage on the continuum of compromise officers engage in and rationalize behavior that just a few years before could not be imagined. At first, acts of criminal commission may appear benign and not terribly different from acts of administrative commission. Evidence that will never be of any use is thrown away instead of being turned in, overtime or payroll records are embellished, needed police equipment is inappropriately purchased with money seized from a drug dealer, expecting "a little something in the envelope" when the officers drop by are but a few examples that officers have easily rationalized. "What the hell, we put our lives on the line and they owe us." A gun not turned into evidence and kept by the officer can become "It's just a doper's; what's the big deal?" The "loyalty versus integrity" dilemma can permit criminal actions to develop into conspiracies . . . whether other officers are actively involved or passively remain loyal and accept what takes place.

Now, the risks are far beyond just administrative reprimands or suspension . . . officers face being fired and criminal sanctions when they are caught. The initially honest, dedicated, above reproach officers now ask, "Where did it all go wrong?" and "How did this happen?" as they face the realities of personal and professional devastation and criminal prosecution. Officers who reach the final stage did not wake up one day and take a quantum leap from being honest, hard working officers to criminal defendants.

Entitlement Versus Accountability

Officers can develop an overwhelming sense of victimization and an intense resentment toward the supervisors and administrators who control their job-role. This can lead to another dilemma . . . a sense of entitlement. Entitlement is a mindset that suggests "We stick together" and "We deserve special treatment." The off-duty officer who is driving 30 mph over the speed limit and weaving in and out of traffic

continued

who tells his passenger, a concerned co-worker, "Relax, I have Mastershields!" implies a sense of entitlement and feeling of impunity. Entitlement allows both on and off duty officers to operate with the belief that many of the rules don't apply to them. "Professional courtesy" goes far beyond just giving another officer a break on a traffic violation. Officers are constantly faced with the dilemma of "doing the right thing" or "doing what they know is right." The only way to change this sense of entitlement is to foster an environment of accountability . . . both organizational and personal accountability.

Loyalty Versus Integrity

Most officers want to be known as loyal and a man or woman of integrity. A problem occurs, however, when a sense of victimization and over-identification with the job sets into motion the dilemma of "loyalty versus integrity" (Mollen Commission, 1994). Here is where officers called in to Internal Affairs and asked questions about another officer lie, many times about a minor issue. When this occurs, the officer has traded his/her integrity for "loyalty" to a fellow officer.

Understanding and Managing Stress

For most people stress at work is a significant concern regardless of their profession. Uncaring or demanding supervisors, unreasonable deadlines, personality conflicts with co-workers, or a heavy workload can create burnout. Whether one is a cop, stylist, or insurance salesperson, this outcome is tragic, both physically and mentally. Many researchers feel that while major occurrences like a divorce or death of a loved one create severe stress, life's little irritations cause the most damage. In the long run, a continual dose of minor annoyances causes the most difficulty. We can all relate to losing things, having to watch our weight, car trouble, or disagreements with co-workers.

The following thoughts can make a difference in dealing with irritations. Make a commitment to yourself to "lighten up" and enjoy life's little pleasures. Discover what irritations are bothering you, then carry out a "plan of action" to correct them.

- Establish long-range goals to put more direction and guidance in your life.
- Be an optimist; expect the best out of life!
- Make a things-to-do list every evening or the first thing in the morning.
- Do not ignore or try to forget about things that upset you. Instead, deal with them; overcome the things that make your life unhappy.
- Enjoy life. Savor the pleasurable, fun stuff that happens each day.

Religious leader Robert Schuller said it best when he talked about what to do when things go wrong. "When faced with a mountain, I will not quit. I will keep on striving until I climb over, find a pass through, tunnel underneath, or simply stay and turn the mountain into a gold mine."

Remember, some of life's greatest pleasures come from the ability to pause and appreciate the things that are really important. Do not get bogged down with self-imposed pressures that, in the long run, have very little meaning or importance. Promise yourself to take more time to watch sunsets, go for walks, have good conversations with your family, and even go fishing.

Medical Conditions

Many physical difficulties are often associated with negative stress, from headaches to serious coronary problems. Most controlled research about the relationship between stress and medical problems of the police has dealt with digestive disorders and heart disease. A 1950 study noted high rates of heart disease mortality among law enforcement officers relative to other occupations. Another documented that 27 percent of the evaluated officers had medium/high- or high-risk coronary disease. Other research conducted during the early 1970s discovered that the police were admitted to hospitals for nonduty matters at a significantly higher rate than other individuals. Almost two-thirds of the concerned officers were admitted for digestive or circulatory difficulties, compared to approximately fifty percent for other occupations.

Nationwide comparison research suggests officers experience ulcers and headaches more frequently than others. Unfortunately, further data showed cops smoke more than the general population. Asthma, thyroid disorders, heartburn, high blood pressure, backaches, and muscle cramps have also been associated with psychological stress.

Many studies generally substantiate that law enforcement careers are related to high instances of serious medical problems. However, there is a lack of data to pinpoint exactly how much more likely officers are to develop medical problems than workers from other occupations. Little data compares evidence of medical disorders in relation to different types of law enforcement agencies; it only overwhelmingly suggests that direct evidence proves a distinct relationship exists between the stress of police work and medical problems. Research must continue until there can be no doubt about the exact relationship.[5]

World-famous author and commentator Paul Harvey has cited law enforcement as "the most dangerous job." He wrote, "Today's law man is under more pressure than any of his predecessors. Individual officers are burdened with increasing legal, societal, and personal obligations, all negative. Dealing with stressful human conflicts everyday, life-and-death decisions, survival decisions, policemen are every day 'wrung out.'"

Harvey further cites, "Routinely, the policeman sees the worst manifestation of human behavior. The sum total of an endless cacophony of such cruelty inevitably eventuates discouragement, depression, despair. Policemen are expected to epitomize manly qualities . . . to be tough, aggressive, dominating, unemotional. Some become gun freaks; others become super jocks, and sometimes super studs. No man can live up to this all the time."

In his commentary Paul Harvey emphasizes the emotional and medical toll of a police career. He refers to the police as "time bombs in blue." He points out that 1,500 cops received psychiatric aid or alcoholism counseling during 1985. Of that number, only 500 officers sought help on their own.

Fortunately, many agencies have or are now contemplating internal psychological counseling units, but only larger departments can afford these divisions. Several years ago the Federal Bureau of Investigation announced that its highest training priority for state and local agencies is the handling of personal stress. Making stress manageable for the individual officer includes an emotional and medical ordeal that is grim, to say the least. Police experience a suicide rate six times higher than the general population. Forty percent of the police do not get help for their problems. Almost half of the forty percent will assault their children or wives. Nearly ten percent of that half will either kill themselves or be killed by a member of their own family.[6]

Paul Harvey was right when he spoke of the inherent danger of being a cop. Contrary to popular belief, stress harms more cops than criminals do. Few, if any, veteran officers have not been wounded by it. Divorce, drug abuse, alcoholism, physical disorders, and suicide are all forms of the misery brought about from stress.

Some view stress as an occupational hazard of law enforcement. As such, it can be reduced or controlled through education and support from a concerned agency, friends, and relatives.

Agency Programs

Progressive police departments began to recognize and deal with stress problems during the 1950s and 1960s. These early attempts were aimed toward alcoholism and often associated with Alcoholics Anonymous. As with much of America's work force, the 1970s saw a dramatic increase of interest toward employee assistance programs by law enforcement.[7]

Is it fair to expect law enforcement agencies to provide effective employee assistance programs? If so, what should it consist of? What can you expect if it is already available? How much should you get involved? If a program is not provided, what can be done to assist in developing one? Yes, it is reasonable to expect your organization to provide an effective, formal program to help in dealing with work-related stress. Furthermore, it is not unreasonable to expect professional services. An employee assistance program usually includes counseling by mental health professionals.

Emotional problems can result from working the street, financial difficulties, postshooting trauma, internal management problems, or an assortment of other family and personal issues. Whether the problems are related to work or family issues, the results are the same. No one goes through a divorce, death of a close friend, alcoholism, or other crises and leaves their troubles completely at home.

Though police departments should provide assistance, effective programs are not inexpensive. The current cost for a program in a medium-size department varies from a couple thousand to nearly twenty thousand dollars annually. Regardless of the amount an agency has funded, there are many programs which may be offered. Several are very beneficial, yet fairly inexpensive. Stress-related programs may include:

- Confidential counseling
- Stress-related needs assessment
- Physical fitness
- Nutritional/weight control
- Spouse training
- Stress management training
- Stress-related supervision training
- Postshooting trauma counseling
- Peer counseling
- Alcoholism counseling
- Drug abuse counseling

No matter what types of stress assistance are offered, they can be organized in several ways. An agency may choose to enter into a contract with an independent counseling agency. Another way is for the counselor to become an employee of the department. A third alternative is some variation of the first two.

Whichever method is chosen, officers must feel that they have an opportunity to voice their opinion in developing the program. A needs assessment should be conducted. Without input, the force may resist its implementation and view stress assistance as something being forced upon them. Everyone must be educated on how the program will operate and absolute confidentiality must be assured.

Crisis intervention, limited counseling, and alcoholism treatment are the most common services provided. Counseling services must be available for officers twenty-four hours a day every day. Marital problems often generate requests for assistance. Though most officers who receive counseling do so voluntarily, a system to provide necessary mandatory referrals is also important. While confidentiality is essential, a monitoring system that documents counseling must be established.

Any good stress or employee assistance program has an element of training. Everyone within the agency should participate in an orientation training class. Training developed specifically for supervisors should focus on their role and responsibilities in stress management. Seminars providing assistance to officers' spouses should be offered on a regular basis. As with all training, stress seminars must be thoroughly developed, based on a needs assessment and performance objectives, presented in a practical manner, and well documented.

Agency Management Style

Most officers agree that the pressures of their jobs can cause serious problems. Many feel that the majority of stress comes from within the department rather than on the street. If a department is riddled with backstabbing, power plays, internal politics, demeaning and ridiculing supervisors, lack of recognition for those who deserve it, and a constant feeling of tension, stress will take its toll.

Law enforcement agencies are structured in a semimilitary management style. Supervision is often authoritarian. Middle- and top-level managers have sometimes risen through the ranks having strict disciplinarian role models. They, in turn, are accustomed to a rigid, disciplined supervision style. Thus, the macho, tough cop image is perpetuated.

In a crisis situation, quick and immediate discipline is essential. There is no time for a debate or discussion. At times like these, authoritarian leadership is crucial. In other ways and at times other than a crisis, a strict disciplinary leadership does little to improve an officer's self-esteem or feeling of achievement. It is past time for law enforcement to change its management style to that of the world's best run corporations. Enlightened leadership is people-oriented. Employees participate in decision making, and an atmosphere of respect, trust, and honesty must replace management ridicule, suspicion, and power struggles.

The ability of an administration to be sensitive to its employees' emotional needs is important. When management ignores the fact that officers need assistance with stress-related problems, they do an injustice to both the organization and its personnel. These situations frequently cause disgruntlement, absenteeism, discipline problems, or apathy. Police departments have been notorious for "not taking care of their own." This, in itself, has been a crime.

Veteran Officer Concerns

As officers age, more than their physical appearance changes. If they have been in the same assignment for years without being promoted, especially the patrol division, feelings of frustration, depression, low self-esteem, and little sense of achievement may have been generated.

One extensive study of veteran officers identified a number of frequent complaints about internal operations:

- Veteran officers are not requested to provide their expertise
- Administrative insensitivity is a common perception of veteran officers
- Days off and shift selection do not take seniority into account
- Promotional processes are perceived to be unfair
- Supervisors show little respect for older officers' experience
- Measurement of productivity is unfair to veteran officers[8]

It is not difficult to understand why senior officers express these views. Most people, given the same situation, would share many of these feelings. It is difficult to understand, though, why some agencies ignore the problems.

Veteran officers should not be left to struggle alone with feelings of despair and frustration. Fighting the pressure from working the street, dealing with the judicial system, and departmental politics are more than enough for anyone to deal with. The administration of any agency should view its role as serving the department by removing obstacles which prevent employees from doing their jobs, not as supervisors who are served by subordinates. Officers, likewise, must be willing to be part of a cooperative team.

Emotional Detachment

He reported that one night during his routine patrol along the waterfront he happened to fall in behind a car being driven by a middle-aged African American man. Both cars drove the same route for several minutes, moving slowly past the industrial buildings, until the car in front suddenly stopped. The driver jumped out and strode back to the police car, which had also halted abruptly. "Why are you following me?" the man demanded angrily. "Can't a black person do anything around here without being harassed and treated like a criminal?"

"While he was raving at me for tailgating him," the officer explained, "I realized that my regular patrol had a whole different meaning for this man. Maybe he had been in trouble in the past; certainly he had been hassled—it sort of goes with the territory of being black in a rich white city. So I let him finish and then said in as offhand a way as I could, 'Do you think I have nothing better to do than follow you around? Why are you being so paranoid?' I added that I hadn't been following him but just happened to be going in the same direction, and he'd gotten there ahead of me. For a minute he didn't seem to believe me—it was like he wanted to stay mad—but he gradually calmed down and muttered something about looking for boxes because he was moving. 'Well,' I said, 'fine, good luck,' and backed up to pull around his car and finish my patrol."[9]

Anyone who has "worked the street" for any length of time can recall similar incidents. Snide remarks by someone an officer has stopped for a traffic citation, a rude business owner having just discovered that his or her business was burglarized, or the cold, frustrated comments of someone who has been a victim of abuse once too often—comments like these take their toll as the months and years pass. Dealing with the emotional turmoil caused from angry and frustrated citizens is easier said than done. It is aggravated by the fact that most people become cops because they are caring individuals; they want to help others.

Remember that good cops can detach themselves emotionally from street frustration. Do not "personalize" rude or angry comments that seem to be directed at you. Whether someone is a mean and nasty person or a good person who happens to be going through a bad time is something over which you have no control. You can, however, control how you react to the situation. Great cops learn to take everything in stride and remain detached.

The alternatives to emotional detachment could be alcoholism, marital problems, drug abuse, departmental discipline, suspension, termination, or suicide. Becoming emotionally involved will cause clouded judgment and decisions, and reacting out of anger or emotion instead of logic and reasoning can be fatal—for you or your career.

Exercise

Besides emotionally separating yourself from the mental roller coaster ride of life as an officer, other things can relieve the effects of stress. Documented, well-researched studies confirm that physical exercise is a great way of releasing stress. Running, racquetball, softball, tennis, swimming, or other sports are more than just fun. Officers must remember that rotating shifts, working weekends and holidays, unhealthy eating, long hours, bad weather, and boredom mixed with rare but intense physical exertion also can be a damaging burden. The benefits of staying healthy and physically fit, therefore, should be obvious to officers; people who stay in sound physical condition know the exhilarating sensation of exercise on their bodies, which also affects mental outlook.

Physical exercise can help relieve some of the negative aspects of stress and keep your body and mind prepared for survival. Imagine what your chances would be if you were overweight and totally out of shape. You have more than an obligation to yourself, and a responsibility to the citizens you are sworn to protect.

Satisfaction from Helping Others

Eat nutritiously, get plenty of exercise, and "be a nice guy"—that is the conclusion that more and more highly acclaimed research is confirming to be sound advice. Continued evidence has shown that helping others also has substantial health benefits. Neuroscientists, epidemiologists, sociologists, and psychiatrists continue to conclude that being of assistance to others provides an inner strength necessary to overcome one's own problems. Furthermore, increased life expectancy and improved vitality have been documented results of giving long-term help to others.

Many studies have indicated that people need meaningful relationships for their own well-being. Nervous disorders, immune system problems, premature death, and deteriorating health have been unproportionately related to people who are unmarried, have few friends or relatives, and do not take part in community activities.

While researchers agree that social involvement provides health benefits, there is still controversy over why or how this occurs. Some researchers believe that a sense of satisfaction from helping others produces endorphins, the brain's natural proteins that induce rest or inaction or quiet uneasiness. This helps the nervous system.

Studies also indicate blood cells comprising the immune system are sensitive to neuropeptides, compounds produced by the brain. Nerve cells connect the brain to various parts of the body. Immune system cells are required to battle infection and are produced in several areas of the body. Though the debate continues, the possibilities are worth remembering and trying yourself.[10]

Disagreements

The proper way to disagree with others is important—whether you are at home or work, with family and friends, or discussing something with your supervisor. Often, needless "hard feelings" are created or friendships severed simply because those involved did not understand the importance of communicating effectively, especially when emotions run high.

The first ingredient of a successful formula for handling disagreements is to disagree without taking it personally. Becoming defensive and angry only escalates the negative aspects of a disagreement. Logical, intelligent people will always disagree on issues from time to time.

It is not wrong to disagree; it is very healthy to have a differing point of view. Do not become defensive. If you take the discussion personally and begin to attack the other person, no one will win. Your use of body language and choice of words can turn a mature, adult conversation into a bitter, childish argument. Remember to put yourself in the other person's place; think how you would feel if you were that person. Understand that the incident which prompted the discussion may not be the real issue. Maybe there is a hidden issue that must be resolved.

Everything does not have to be resolved during the disagreement. If emotions begin to flare, calm down. Agree that the discussion can be continued later on in the day or another day. Thinking things through has a way of resolving delicate situations. In addition, one of you may simply have had a rotten day, and the entire situation may not be as gloomy as it appears at the moment.

Lastly, a few techniques may help. First, talking softly eases heated situations; it is very difficult to argue with someone who is calm and soft-spoken. Second, don't take "cheap shots" at the other person. Intentionally hurting someone's feelings does nothing but hurt everyone involved. You certainly won't feel good about yourself. Third, it is important to be able to forgive and forget. To harbor a grudge is of no benefit to anyone. Life is too short to handicap yourself emotionally.

"A great cop is like a well trained athlete. He is a competitor who always wants to win, hates to lose, but believes in the game and plays by its rules. He loves everything about policing, from providing service to literally fighting with criminals."

Michael Sanford
Administrative Aide to Chief of Police
Seattle Police Department

PERSONAL RELATIONSHIPS

From our childhood years until we become senior citizens, we have many relationships in our professional and personal lives. The ability to balance these relationships with our many other commitments can be challenging, especially in today's multitask, multirole society. Meeting that challenge is important to our well-being, happiness, and self-satisfaction.

Some of the most important relationships we have are those with spouses, our children, and significant others. Keeping these relationships alive and well requires commitment from all parties involved, honesty, respect, communication, and, of course, love.

Spouses

Most people know that half of all marriages in the United States end in divorce. The divorce rate for police officers is even higher. Why are cops such poor risks when it comes to marriage?

Sometimes a spouse of an officer feels as though police work takes priority over the couple's relationship. Law enforcement has a way of becoming an obsession with some new officers. Veteran officers can look around the department and point out rookies who literally eat, breathe, and live for their jobs.

Obsession with police work, though not uncommon during the first few years on the force, is not healthy. Some officers work thirty to forty hours off duty, without pay, every week. They would not consider being out sick and sometimes do not even want to take vacations. Even spending hour after hour around the station talking shop is common. Their husbands or wives would at first understand. Later they may feel hurt and rejected.

Rotating shifts and unpredictable work hours are more causes of marital strain. They lead to nights sleeping alone, lack of sleep, and more marital stress.

Another reason for marital stress is the perceived or real sexual temptation that many officers face. Few spouses are comfortable with the thought of their better half driving around all night with someone of the opposite sex. Inevitably there are stories of men and women who throw themselves at officers in uniform.

Officers' spouses often worry about the danger their loved ones face. The caring must be sincere and the commitment strong. It is no secret that marriage takes effort.

The most valuable asset to any police marriage is having the commitment to work through everyday problems. Take holidays as an example. Uniform officers usually have to work on holidays. Changing work schedules to make your immediate family and relatives happy is not an easy task; everyone must learn how to compromise. Visiting several relatives at their homes may be necessary. Even an awkward schedule may be required. In any event, the key is to put yourself in your spouse's position.

Another ingredient for success is the ability to communicate. Once this art is lost you are both in trouble. Do whatever it takes to keep open, honest and respectful communication alive. Make it a point to create situations that encourage talking. Go for walks. Go fishing, swimming, hiking, to the ball game, or aerobics class together. Communicating is more than talking; it is body language and just being there for the other person. Remember, the best part of life is the here and now, so share it together.

Understanding and compassion are important qualities. You are responsible for sharing and caring about your spouse's feelings. People worry because they care, and you must help your spouse through the rough moments. It is frustrating going to bed alone, having holidays disrupted, and feeling the insecurity of knowing sexual temptation may surround your mate. It is not difficult to understand how they feel. For the most part, you are the only one in a position to help. Officers have to show understanding and compassion for the problems their jobs create in their marriages. A defensive attitude only takes them further down the road to the marital war zone, and that is a battle no one wins. Don't fight it.

Force yourself not to become obsessed with work; you can still be a great cop. Even better, you will be a balanced person. Take time out. Go for minivacations whenever possible. Talk with each other. Enjoy life. Reach a healthy balance between work and your personal life. Otherwise, your marriage and career will suffer.

A happy marriage means you are more likely to be a better cop. It takes commitment, understanding, caring, and a lot of give and take.

Kind Words

Words of kindness between husband and wife help to keep a marriage from growing cold and bitter. Like most aspects of marriage, constant attention must be given to how and what we say to each other.

Time has a way of causing couples to take each other for granted. Sometimes people tend to be overly critical, and destructive criticism and verbal jabs are very damaging. To make matters worse, such a cycle feeds on itself, becoming only more vicious with time. On the other hand, if a couple makes a habit of giving praise, just the opposite is true. Warm, sincere compliments are priceless; they cost nothing and are invaluable in times of sadness. In fact, few things in life are as effective as a sincere smile and a kind remark.

People who have grown up in households where words of kindness were rare are unlikely to feel comfortable giving or receiving compliments. They simply are unaccustomed to it, leading to awkwardness or outright feelings of defensiveness. These feelings can overwhelm people who believe their spouse is pressuring them to pay more compliments.

These suggestions may help:

- Use a variety of ways to show your love and appreciation.
- Be creative.
- Give praise about the things that are most meaningful.
- Be spontaneous. Do not let yourself get in a rut.
- Be sincere in your compliments. Do not say things that are not true.

- Remember that the simplest deeds and remarks often mean the most.
- Romance can be a rewarding part of life.
- Express how much you enjoy your spouse's company. Learn to appreciate being with each other more and more, and tell each other how you feel.[11]

Significant Others

Much of the information presented for spouses also applies to officers considering serious relationship commitments or marriage. Like spouses, significant others, too, can find it difficult to understand or accept why officers must work long hours, miss some holidays, or rotate work schedules. They also worry when their loved ones face danger every day or meet other temptations. As an officer, you can be sensitive to your partner's concerns by communicating openly and honestly and having compassion for him or her.

In any serious love relationship, both partners need to clearly understand and communicate their level of commitment to one another. Perhaps one or both of you is ready for a further commitment. If so, it is wise before taking that next step for both of you to understand the type of work that you, as an officer, will be involved in and the commitment you must give to your job and to each other. A balance between your relationship and your work is possible with effort and understanding from both partners.

Children

As if growing up isn't hard enough, some kids are the children of police officers. Children of anyone in a high-visibility occupation occasionally will reap the benefits or burdens of their inherited situation. They need to be emotionally prepared to handle these situations, and parents are responsible for seeing that their children are emotionally ready for the future.

Every child looks to his parents for nurturing, advice, security, and role modeling. Keep an open line of communication. Disagreements will be inevitable, especially as children mature into young adults, because they begin to resist authority and strive for independence. You do not have to win every disagreement; if you feel you do, children will rebel. Have compassion for the emotional tornado they will experience during puberty and other stages of growing up.

Parenting is one of life's most difficult, but rewarding, challenges. Warmth, frustration, love, anger, caring, and endless memories are all part of being a parent. Some aspects of police work can make parenting an even greater challenge—long hours, missed holidays, potential danger—but do not miss the wonderful joys of raising children. Spend more time with them. Go fishing together, toss a ball around, go bowling, help them with homework, or simply sit and talk. You will cherish these times forever.

O, GOD

"Give me wisdom of these that criticize me, so that I might not make wrong decisions again.
Give me patience with the young.
Give me compassion and understanding with the elderly.
Give me the wisdom to remain silent when I would rather express my thoughts loudly.
Give me the determination to keep on trying when plans fail, leads run out and information obtained is erroneous and all seems lost.
When the opinions of others differ from mine, let me listen with an open mind and consider the possible alternative.
Give me the humility to admit when I'm in error and the wisdom to stay silent when I'm right. Let me speak up for what is right and just.
Help me view and treat all persons as an equal, regardless of race, color, creed, sex or station in life.
Give me the courage to perform the duties of my job, even in the face of great danger or performing a task that is unpopular.
Let me put the safety of others ahead of my own safety.
When my job is finished and my head is hanging low, give me something else to do so I might show my love for you."

By an Old Lawman, Just a Cop to
Quite a Few
In *Police Stress,* Edward C. Donovan, Editor,
The International Law Enforcement Association, Mattapan, MA
Winter 1987

End Notes

1. Gail Sheehy, *Passages,* Bantam Books, New York, 1976, pp. 39–46.
2. M. Steven Meagher and Nancy Yentes, "Choosing a Career In Policing: A Comparison of Male and Female Perceptions,"(paper presented to the Annual Meeting of the Academy of Criminal Justice Sciences, March 20, 1986, Orlando, Florida)
3. John G. Stratton, "Police Stress: An Overview," *The Police Chief,* Gaithersburg, MD, 1978, pp. 58–61.
4. Robert A. Jud, "Making Stress Manageable," *Business Week,* July 1987, pp. 75–76.
5. Gail A. Goolkasian, Ronald W. Geddes, and William DeJong, *Coping With Police Stress,* U.S. Department of Justice, National Institute of Justice, U.S. Government Printing Office, Washington, DC, 1985, pp. 5–8.
6. Paul Harvey, "The Most Dangerous Job: Law Enforcement," *The Los Angeles Times Syndicate,* Los Angeles, California, 1986.
7. Gail A. Goolkasian, Ronald W. Geddes, and William DeJong, p. 12.

8. Mark Pogrebin, "Alienation Among Veteran Police Officers," *The Police Chief,* Gaithersburg, MD, 1987, pp. 38–40.
9. Edward E. Shev and Jeremy Joan Hewes, *Good Cops/Bad Cops,* San Francisco Book Co., 1977, p. 98.
10. Eileen Rockefeller Growald and Allan Luks, "The Good Guys Finish Healthier, Research Shows," *The Orlando Sentinel,* Orlando, FL, October 4, 1988, p. E–3.
11. Diane Hales, "Words That Can Warm Up Your Marriage," *McCalls Magazine,* April, 1989, pp. 70–72.

CHAPTER 7

STAYING PHYSICALLY FIT

"Being a great cop is about being ethical and being a professional. It is about being a leader and not a follower, and setting an example as a highly skilled and dedicated individual. A great cop sets ethical standards others will want to emulate."

John Linn, Ph. D. Penn State University
Criminal Justice, Altoona College

In the last couple of decades the United States has witnessed a major shift toward physical fitness. We have been bombarded with exercise—television shows, magazines, books, spas, health clubs, aerobics, prayercise, and more.

Most fitness information endorses a particular diet or exercise program. With the exception of diet fads, almost everything from a reputable source has merit. All fitness programs should begin with a physician's examination.

IMPORTANCE OF FITNESS

Americans lead the world in deaths from cardiovascular-respiratory difficulties. Staying physically fit is, literally, a matter of life and death. Understanding and appreciating the principles of cardiovascular health is not difficult, and it would be foolish not to do something to prevent heart disease. Most people exercise for appearance, health, and enjoyment. The following are heart disease risk factors that can be reduced through exercise:

- Obesity
- High blood pressure
- High blood sugar levels
- Stress
- High blood fats levels
- Cardiovascular inactivity

Other major risk factors for heart disease are cigarette smoking and some inherited conditions.

Exercise allows you to live a more enjoyable, productive life. It helps you feel better, have more energy, and be much more resistant to injury. It relieves tension, improves attitude and increases self-esteem naturally. Specifically, staying physically fit can:

- Result in more efficient use of the lungs
- Develop a stronger, more efficient heart
- Create improved digestion and bowel movement
- Prevent lower back pain
- Reduce body fat
- Improve posture
- Reduce blood pressure
- Improve leg circulation
- Reduce the negative aspects of stress
- Enhance self-confidence
- Reduce health expenses
- Increase oxygen consumption
- Reduce resting pulse rate
- Improve athletic performance
- Enhance career achievement through role modeling and appearance
- Improve family relationships[1]

Imagine pursuing someone on foot for several blocks then being assaulted by one or two suspects. Consider that you are 25 pounds overweight and never exercise, and that they are athletic and fit. Imagine your numbness and exhaustion.

Make a commitment to stay in shape. Remember that you are not the only one who depends on your physical conditioning. Time and time again cops count on each other, and being unfit can get a fellow officer killed.

YOUR FITNESS LIFESTYLE

Fitness should be viewed as a continual lifestyle, not a program or diet. A commitment to good health must be lifelong, not short term. It should be based upon common sense and reputable research. Fad diets are often not worthwhile. Height/weight tables are frequently misleading. In addition, there is no need to buy a lot of athletic clothing or join an expensive spa. Anyone with a dedicated attitude will find the necessary requirements simple and straightforward.

Establish Goals and Objectives

Decide what you want to accomplish, and then establish meaningful goals. Perhaps you want to lose a specific amount of weight, increase strength or endurance, improve cardiovascular activity, or lower cholesterol levels. Once you establish goals, set specific objectives. For example, if your overall goal is to lose 40 pounds within one year, the objective might be to lose 3–4 pounds each month. Objectives can show your progress and motivate you to continue in pursuit of a goal.[2]

Physical Examination and Assessment

Begin your evaluation with a physical examination and assessment by a physician. This will help you avoid overworking yourself and endangering your health unknowingly. Tell the physician that you intend to develop a fitness lifestyle. He or she should give special attention to your circulatory and heart condition, cholesterol levels, and joint and skeletal conditions. In addition, the physician should evaluate your fitness lifestyle in terms of medical conditions you may have, such as diabetes or back trouble. The exam may also alert you to previously undetected medical conditions or problems.

After a physician explains the results of the exam, reassess your lifestyle. Do you smoke or drink? Are you overweight? What kind of and how much exercise do you do? How much stress are you under? Are your eating habits of concern?

Relate the answers to the goals and objectives you established. Then decide whether lifestyle changes are necessary to meet and sustain your goals.

Assess Your Physical Fitness

An objective evaluation of your physical fitness can be eye-opening. Once the physician's exam is complete, appraise your abdominal conditioning, upper body strength, and cardiovascular system.

Abdominal Conditioning To assess abdominal conditioning, do as many sit-ups as you can in two minutes without stopping.

1. Lie flat on your back with knees bent, feet held down on the floor by a partner.
2. With your arms straight out in front of your body, raise yourself forward until your elbows reach your knees.

3. Lower yourself back to the floor.
4. Repeat the entire process. If you stop for more than a second at any point, the test is over.

You should be able to do at least 30 sit-ups in the allotted time without stopping.

Upper Body Strength To assess your upper body strength, do as many push-ups as you can within two minutes.

1. Assume the standard push-up position: body and head nearly parallel to the floor, arms almost straight up and placed at a width equal to your shoulders, legs together and propped up on toes. Ensure that your body is as straight as possible.
2. Lower your body to within three inches of the ground. Keep your entire body totally straight during the test.
3. Raise your body until arms are straight again.
4. Repeat the entire process. If you stop for more than a second at any point, the test is over.

You should be able to do at least twenty-five push-ups.

Cardiovascular Condition The most important element of fitness is the cardiovascular system. To a street cop, muscular strength is important, as is the ability to sustain physical exertion without fatigue. The key to superior endurance is a cardiovascular system developed to the extent that it supplies enough oxygen to muscles in motion while improving their ability to use additional oxygen. An officer who is in poor condition will produce an excess of the chemical lactic acid, which tends to induce fatigue, which destroys the ability to defend oneself.

Measuring your heart's pulse rate immediately after exercise is a good way to check cardiovascular fitness. Place the index and the middle fingers together on your neck beside the Adam's apple and count the number of beats you feel for thirty seconds, then multiply the result by two. This is your resting heart rate. Most people have a rate between 50 and 90 beats a minute. A lower rate usually means good cardiovascular conditioning.

To assess cardiovascular fitness, do the step test for three minutes.

1. Use a stool or a similar sturdy object that is 8 inches high. Stand in front of the stool and step up and down with both feet every five seconds, alternating the first foot each time.
2. Thirty seconds after stopping, take your pulse rate for 30 seconds, then multiply that number by two.

An individual in good cardiovascular condition can expect to have a pulse rate in the 120s. Someone in poor cardiovascular condition will have a rate in the 160s or 170s, generally speaking.

These tests should give you a fairly good idea of your physical condition. Begin a fitness diary; write the date and scores of these three tests.

Each time you take the tests, record the same information. It is motivating and fun to track your improvement.[3]

Choose Your Exercise

Like the general public, some officers get virtually no exercise. For the most part, it is because they just do not enjoy it. Many people think of exercise as something that causes them pain and anguish and is not much fun. Most of these people push themselves too hard while beginning. The key to remember is that if exercising hurts, slow down! Do not try to get in condition in a couple of weeks. The results could be dangerous to your health.

Understand discomfort before beginning. The basic guideline is do not ignore discomfort or pain, particularly if it is in the chest or left arm or if you have a sensation of numbness or dizziness. Pain in these areas could be a sign of heart difficulty or attack. Also take into account high temperatures. Seek medical assistance immediately if you have a high temperature or numbness or dizziness in your chest or left arm.

Keep in mind these additional factors:

- If you have stopped exercising for more than a week, begin at a reduced level.
- Be patient—getting in shape takes time.
- Do not exercise on a full stomach.
- During warm months, drink 6–8 ounces of fluid before exercising.[4]

There are many ways to get and stay in shape, so choose something you like; otherwise, you will eventually quit. Some people have more fun participating in team sports while others prefer solitary pursuits such as running, hiking, or cycling. Remember to consider the goals that you initially set. Some exercises will help you reach your goals better than others.

Consider also the cost of equipment. If you are faced with not having enough money to select your favorite sport or exercise, start with an exercise or sport you can afford. Do not use lack of money as an excuse for inactivity.

STRETCHING

The first step of every workout should be stretching, also called the warm up. It is the vital transition between a sedentary and active state that helps to prepare you for exertion, keeps muscles supple, and prevents stress and strain throughout the body.

Some exercise, like bicycling, racquetball, or running, generates particular muscle tightness. Stretching before and after each workout affords maximum flexibility of your muscles and prevents injuries. Achilles tendonitis, shin splints, and pulled muscles can be very painful. When these types of injuries occur, you may have to face the frustration of having to get in shape all over again.

Stretching is easy, but many people do not know the proper techniques. If it is done incorrectly, it can do more harm than if it were never done at all. The basic rule is to remember that stretching feels good when it is done properly and hurts if it is done improperly. Relax when you stretch and establish a regular stretching routine. If you find that the stretched muscles hurt, you have gone too far and should reduce the stretch.

Stretching should be enjoyable, because it relaxes and energizes. Remember, no one gets in shape within a few days. Long-term commitment, patience, and a good attitude will make your workout a pleasure.

Stretching should become a habit for several reasons:

- It increases blood circulation.
- It improves your range of motion.
- It makes you feel good.
- It improves your awareness of different body areas.
- It decreases the possibility of injury.
- It enhances coordination.
- It reduces tension and stress.
- It helps to give you confidence and self-esteem.

Convenience

Anyone can enjoy stretching. It makes no difference if you work in an office, out of a squad car, or ride a motorcycle. Unlike many types of exercise, stretching can be done whenever you want: at your desk, walking through a parking lot, at home, even while in a car. In addition to stretching before and after exercise, there are several other times when it is particularly helpful.

- Whenever you feel tight or stiff
- During or after standing or sitting for long periods
- To release nervous tension
- As soon as you get up to begin the day
- To improve a tension headache

Technique

Bouncing and stretching until it hurts is an incorrect stretching technique. Stretching should be a sustained, relaxed process. Loosening up tense muscles cannot be accomplished quickly. It becomes easier after time if done properly and on a regular basis.

Hold each muscle stretch for 10 to 30 seconds. Nationally recognized stretching and fitness authority Bob Anderson describes the sensation you should feel as a "mild tension." He goes on to say, "The feeling of tension should subside as you hold the position. If it does not, ease off slightly and find a degree of tension that is comfortable." Never bounce!

Breathing should be slow and systematic. Holding your breath is not recommended. If you are unable to breathe slowly and naturally, you should relax more. Take everything slow and easy and teach yourself to breathe naturally.[5]

Use a book for reference that illustrates stretching in detail. The text *Stretching* by Bob Anderson is highly recommended. It is published by Shelter Publications, Bolinas, California.

"A great cop is an honest person that has a sense of humor, a lot of common sense, and the courage to use it."

J. E. Tillman
Chief
North Las Vegas Police Department

NUTRITION AND EATING HABITS

Contrary to popular belief, dieting is not recommended, even for those with a weight problem. Diets are usually meant to be temporary. After most people complete a diet, they usually put the weight back on, sometimes more weight than they started with.

The best method for weight control and a healthy body is nutritional eating habits and a healthy lifestyle. Frustration and disappointment are inevitable unless the right nutrition, eating habits, and exercise exist. The proper blend of these makes a healthier, happier life possible and allows your body to maintain, repair, and develop itself.

In addition to the performance we all seek, preventing illness is a major goal. People usually do not appreciate good health until they no longer have it. Following nutritional guidelines will help to prevent many things, from a heart attack to digestive trouble. Maintenance of the body is similar to that of a well-operating vehicle. Investing the relatively small price of exercising and eating right will prevent you from having to pay the devastating price of losing your health.

Most of us know about the four basic food groups referred to in Table 7.1. These food groups comprise the foundation upon which our good health rests; a strong foundation usually means a strong body, too.

Eating the right amount of each group will furnish sufficient basic nutrition; it does not involve a complicated formula. Its simplicity, however, does not equal its extreme value.

TABLE 7.1
BASIC FOOD GROUPS

Food Group	Servings
Grain and Cereal Flour, whole-grain foods, baked items, enriched bread	4 servings per day
Fruit and Vegetables Includes a wide assortment of vegetables and fruits	4 servings per day
Dairy Products Cheese, yogurt, milk, and milk products	2 servings per day
Meat and High-Protein Items Fish, poultry, meat, nuts, beans, and other high-protein foods.	2 servings per day

The information in Figure 7.1 and Table 7.2 provides additional guidelines on nutritional value, food consumption, and calories. It will help you make more informed eating and exercise choices.

FIGURE 7.1

From the Office of the Minnesota Attorney General

Did you know there are more than 300,000 fast food restaurants in the U.S.? Why is fast food so popular? Because it is convenient, predictable, and fast. Fast food has become a part of the busy American lifestyle. Yet, as nutrition experts point out, fast food is often high in calories, sodium, fat and cholesterol. This does not mean fast food is bad. But it does mean you should fit fast food into a balanced, healthy diet.

To help you make fast food choices and be an informed consumer, the Minnesota Attorney General's Office has developed the guide *Fast Food Facts,* which the Food Finder is derived from. Included are the calorie, fat, sodium and cholesterol counts of menu items from popular fast food restaurants, based on the companies' own nutritional analyses. Below are some basic facts to help you make nutritional comparisons with this guide.

Calories

On the average, to maintain desirable weight, men need about 2,700 calories per day and women need about 2,000 calories per day. It is not well understood why some people can eat much more than others and still maintain a desirable weight. However, one thing is certain, to lose weight, you must take in fewer calories than you burn. This means that you must either choose foods with fewer calories or you must increase your physical activity, preferably both.

Fat

Research shows that eating too many high-fat foods contributes to high blood cholesterol levels. This can cause hardening of the arteries, coronary heart disease and stroke. High-fat diets may also contribute to a greater risk for some types of cancer, particularly cancers of the breast and colon.

While most Americans get more than 40 percent of their daily calories from fat, the American Heart Association recommends limiting fat to less than 30 percent of daily calories. This means limiting the fat you consume to 50–80 grams per day.

Percent of Calories from Fat

The category in this guide "% of Calories from Fat" is calculated by multiplying the grams of fat by nine (there are nine calories per gram of fat), then dividing the calories of fat by the total number of calories in the food.

Cholesterol

The American Heart Association recommends eating no more than 300 milligrams of cholesterol per day. But don't just look at the cholesterol contained in a food item. A product high in total fat or saturated fat can be an even bigger contributor to high blood cholesterol levels. For example, "cholesterol free" potato chips may be high in fat and may contribute to raising your cholesterol level, because high-fat foods cause the formation of cholesterol in the body, even if the food itself contains no cholesterol.

Salt

Everyone needs some sodium in the diet to replace routine losses. The Food and Nutrition Board of the National Academy of Sciences/National Research Council has estimated that an "adequate and safe" intake of sodium for healthy adults is 1,100 to 3,300 milligrams a day, the equivalent of approximately 1/2 to 1 1/2 teaspoons of salt. Americans, on average, consume at least twice that amount, 2,300 to 6,900 milligrams of sodium daily, according to estimates by the Food and Nutrition Board. For some people, consuming high amounts of sodium can cause high blood pressure.

Fast Food Meals

Fast food meals can be high in calories, fat, sodium, and cholesterol. See how easily these red-flag items can add up:

1. Burger: Quarter-Pound Cheeseburger, Large Fries, 16 oz. soda *(McDonald's)*

This meal:	*Recommended daily intake:*
1,166 calories	2,000–2,700 calories
51 g fat	No more than 50–80 g
95 mg cholesterol	No more than 300 mg
1,450 mg sodium	No more than 1,100–3,300 mg

2. Pizza: 4 Slices Sausage and Mushroom Pizza, 16 oz. soda (*Domino's*)

This meal:	*Recommended daily intake:*
1,000 calories	2,000–2,700 calories
28 g fat	No more than 50–80 g

continued

62 mg cholesterol	No more than 300 mg
2,302 mg sodium	No more than 1,100–3,300 mg

3. Chicken: 2 Pieces Fried Chicken (Breast and Wing), Buttermilk Biscuit, Mashed Potatoes and Gravy, Corn-on-the-Cob, 16 oz. soda (*KFC*)

This meal:	*Recommended daily intake:*
1,232 calories	2,000–2,700 calories
57 g fat	No more than 50–80 g
157 mg cholesterol	No more than 300 mg
2,276 mg sodium	No more than 1,100–3,300 mg

4. Taco: Taco Salad, 16 oz. soda (*Taco Bell*)

This meal:	*Recommended daily intake:*
1,057 calories	2,000–2,700 calories
55 g fat	No more than 50–80 g
80 mg cholesterol	No more than 300 mg
1,620 mg sodium	No more than 1,100–3,300 mg

Better Fast Food Choices

This brochure is not meant to scare you away from fast food entirely. Rather, it is intended to provide you with information to help you make better fast food choices. Realize that it is still possible to eat fast food occasionally and follow a sensible diet. See how these meals stack up against the previous examples:

1. Burger: Hamburger, Small Fries, 16 oz. soda (*McDonald's*)

This meal:	*Recommended daily intake:*
481 calories	2,000–2,700 calories
19 g fat	No more than 50–80 g
30 mg cholesterol	No more than 300 mg
665 mg sodium	No more than 1,100–3,300 mg

2. Pizza: 3 Slices Cheese Pizza, 16 oz. diet soda (*Domino's*)

This meal:	*Recommended daily intake:*
516 calories	2,000–2,700 calories
15 g fat	No more than 50–80 g
29 mg cholesterol	No more than 300 mg
1,470 mg sodium	No more than 1,100–3,300 mg

3. Chicken: 1 Piece Fried Chicken (Wing), Mashed Potatoes and Gravy, Cole Slaw, 16 oz. diet soda (*KFC*)

This meal:	*Recommended daily intake:*
373 calories	2,000–2,700 calories
19 g fat	No more than 50–80 g
46 mg cholesterol	No more than 300 mg
943 mg sodium	No more than 1,100–3,300 mg

4. Taco: Three Light Tacos, 16 oz. diet soda (*Taco Bell*)

This meal:	*Recommended daily intake:*
420 calories	2,000–2,700 calories
15 g fat	No more than 50–80 g
60 mg cholesterol	No more than 300 mg
840 mg sodium	No more than 1,100–3,300 mg

Fast Food Surprises

Fast-food chains have noticed that consumers are more health-conscious, and as a result many chains are adding healthier choices to their menus. Here are examples of some of these better alternatives:

Arby's Light Roast Chicken Sandwich

276 calories
7 g fat
23% calories from fat
33 mg cholesterol
777 mg sodium

Burger King's Chunky Chicken Salad

142 calories
4 g fat
25% calories from fat
49 mg cholesterol
443 mg sodium

McDonald's Vanilla Shake

310 calories
5 g fat
15% calories from fat
25 mg cholesterol
170 mg sodium

Wendy's Chili

210 calories
7 g fat
30% calories from fat
30 mg cholesterol
800 mg sodium

Eating For Athletic Competition

While good eating habits are the same for anyone, there are a few additional guidelines for those who exercise intensely. Endurance levels of swimming, jogging or cycling may deplete the necessary supply of carbohydrates during and after strenuous workouts. As a result, increase your intake of carbohydrates.

Adjusting carbohydrates and protein for an increase in exercise makes good sense. Making sure you have enough fiber and controlling cholesterol is good advice for anyone. Besides, fresh fruit and vegetables taste good, even if you aren't used to them.

This type of advice takes us beyond the basic foundation of nutrition to a slightly more specific level. Don't settle on the relatively brief section of this chapter in your quest for better health. My intent has been to merely whet your appetite enough to spur on continued learning about fitness. Many respected fitness books are available at your nearest library, bookstore or Amazon.com.

continued

All 'Round Great Foods

The following is an "everything considered" list of healthy foods. For ease of review, the foods are listed by food groups. All have high nutritional value, with little detrimental aspects.

Fruits
Watermelon
Papaya
Cantaloupe
Mango
Orange
Grapefruit

Vegetables
Spinach
Collard Greens
Sweet Potato
Chickpeas
Navy Beans
Potato
Kale
Squash
Lentils
Split Peas

Beverages
Carrot Juice
Orange Juice
Tomato Juice
Grapefruit Juice
V-8 Juice
Apple Juice

Grain Foods
Cracked Wheat
Whole Wheat Bread
Brown Rice
Rye Bread
Enriched White Bread
Enriched White Rice

Poultry, Fish, Meat, and Eggs
Tuna
Lobster
Roasted Chicken
Salmon
Turkey
Flounder

Dairy Food
Low Fat, Plain Yogurt
1% Fat Cottage Cheese
1% Fat Buttermilk
2% Fat Milk
Skim Milk
1% Fat Milk

Snacks
Carrots
Green Peppers
Dried Apricots
Raisins
Apples
Bananas

Source: J. Boyce Davis and E. Leslie Knight, *CVR Fitness: A Basic Guide For Cardiovascular Respiratory Exercise,* Kendall Hunt Publishing Company, Dubuque, Iowa, 1976.

Table 7.2
Burning Calories

Activity	*Calories (Burned in 30 minutes)*
Aerobic Dance Class	178
Bicycling: 10 mph, on level ground	258
Bowling	108
Fishing	124
Golf	
Carrying clubs	162
Using power cart	108
Handball	
Leisurely	270
Vigorously	297
Horseback Riding	
Galloping	255
Trotting	204
Walking	75
Ice Skating	
Fast pace	315
Slow pace	199
Jump Rope	223
Rowing Machine	378
Roller Skating	
Fast pace	315
Slow pace	199
Running	
5.5 mph	295
7.5 mph	426
Skiing	
Cross-country: 2.5 mph	252
Downhill	247
Sleeping	32
Squash	393
Swimming	
Breaststroke: 1 mph	300
Crawl, fast	291
Backstroke	315
Tennis	
Singles	216
Doubles	162
Volleyball	93
Walking 3 to 3.5 mph	130

Note: The approximate number of calories burned is based upon a 135-pound woman. Increase the number by 10% for every 15 pounds over 135. Decrease this amount by 10% for every 15 pounds you are under 135 pounds.

Source: Trinity Medical Center, Carrollton, Texas, 1989.

STAY MOTIVATED

If you are not staying in shape, check your attitude; it could be the only thing preventing you from having better fitness. People that are not physically fit usually lack the motivation necessary to live a healthy lifestyle. For the officer who refuses to stay in shape, a lack of commitment can mean someone may be hurt because the officer did not have the physical abilities required in the moment.

Consider these aspects:

- You may not be the strongest or fastest, but you can make the most of your natural abilities. Making the best of your capabilities is one of life's greatest responsibilities and satisfactions.
- Imagine the arteries of your body literally clearing each time you exercise.
- Whether your goal is losing weight or reaching new athletic heights, there will be plateaus, times when it appears you are making little progress. Your body is catching up with the change it is undergoing. View these periods as merely stepping stones toward better things to come.
- Pick up a 10-pound weight and carry it around for 20 minutes. Imagine how much easier everyday life will be once you have lost those ten pounds.
- Think about the greatest accomplishments that have been achieved throughout history. Ask yourself what your greatest achievements have been. All great endeavors take sacrifice and determination. Having a healthy body, good marriage, or great law enforcement career all demand an enormous amount of hard work and commitment, but they are worth it.
- Remember, every time you work out and eat right, you are increasing the length of your life.
- Think about how much more confident you will be with your new, toned body.
- Consider what it will feel like to get back into your favorite pair of pants, shirt, skirt, or blouse.[6]

The best commitment you can make in becoming physically fit is to believe in yourself. The rewards will be invaluable, and your body will appreciate the special treatment it deserves.

END NOTES

1. J. Boyce Davis and E. Leslie Knight, *CVR Fitness,* Kendall/Hunt Publishing Co., Dubuque, IA, pp. 1–27.
2. Frank I. Katch, William D. McArdle, and Brian Richard Boylan, *Getting In Shape,* Houghton Mifflin Co., Boston, MA, 1979, pp. 1–8.
3. Ibid. pp. 33–40.
4. J. Boyce Davis and E. Leslie Knight, pp. 31–51.
5. Bob Anderson, *Stretching,* Shelter Publications, Inc., Bolinas, CA, 1984, pp. 8–12
6. Bob Wolff, "Motivation," *Muscle and Fitness,* Escondido, CA, August 1989, pp. 177–78.

To be a great cop, you must have a strong work ethic, a commitment to doing what is right and have a concern for the preservation of mankind.

Your integrity must be beyond reproach, for without integrity you cannot be a law enforcement officer.

You have to be self-initiated and have a strong sense of self-esteem.

You must be honest and fair without regard to your personal prejudices and model this value for all people.

You must have courage and the conviction to act when called upon.

Prepare for anything and know you will persevere.

Fight to win and value life above all.

You must cherish your family.

You must honor spiritual values.

You must be gracious.

You must be patient.

Be educated and informed.

You must embrace laughter *and* tears.

Corinne Garrett
Lieutenant, Orange County Sheriff's Office
Orlando, Florida

APPENDIX A

FEDERAL LAW ENFORCEMENT AGENCIES

Bureau of Alcohol, Tobacco and Firearms
650 Massachusetts Ave., N.W.
Washington, D.C. 20226
(202) 927-8700 Fax (202) 927-8876
www.atf.treas.gov

Federal Bureau of Investigation
J. Edgar Hoover Building
Washington, D.C. 20535
(202) 324-3000 Fax (202) 324-4705
www.fbi.gov

Drug Enforcement Administration
Washington, D.C. 20537
(202) 307-8000 Fax (202) 307-7335
www.usdoj.gov/dea

Financial Crime Enforcement Network
2070 Chain Bridge Rd.
Vienna, VA 22182
(703) 905-3591 Fax (703) 905-3690
www.ustreas.gov/fincen

National Park Service—U.S. Park Police
Ohio Dr. S.W.
Washington, D.C. 20242
(202) 619-7350 Fax (202) 205-7981
www.Doi.gov/u.s. park.police

U.S. Customs Service
1300 Pennsylvania Ave., N.W., Suite 44A
Washington, D.C. 20229
(202) 927-1010 Fax (202) 927-1380
www.customs.treas.gov

U.S. Department of Treasury
1500 Pennsylvania Ave., N.W.
Washington, D.C. 20220
(202) 622-1100 Fax (202) 622-0073
www.ustreas.gov

U.S. Fish and Wildlife Service
4401 N. Fairfax Dr.
Arlington, VA 22203
(703) 358-1949 Fax (703) 358-2271
www.fws.gov

U.S. Marshals Service
600 Army Navy Dr.
Arlington, VA 22202
(202) 307-9001 Fax (202) 557-9788
www.usdoj-gov/marshals

APPENDIX B

LAW ENFORCEMENT TRAINING ORGANIZATIONS

American Society of Law Enforcement Trainers (ASLET)
P.O. Box 361
Lewes, DE 19958
(302) 645-4080 Fax (302) 645-4084
www.aslet.com

Association of Public Safety Communications Officials International, Inc.
2040 S. Ridgeway Ave.
South Daytona, FL 32119
(904) 322-2500 Fax (904) 322-2501
www.apcointl.org

Canada Association of Chiefs of Police (CACP)
130 Albert St., Suite 1710
Ottawa, Canada K1P 5G4
(613) 233-1106 Fax (613) 233-6960
www.cacp.org

Federal Law Enforcement Training Center
Glynco, GA 31524
(912) 267-2224 Fax (912) 267-2495
www.fas.org

Institute of Police Technology and Management
12000 Alumni Dr.
Jacksonville, FL 32224
(904) 620-4786 Fax (904) 620-2453
www.unf.edu

International Association of Arson
300 S. Broadway, Suite 100
St. Louis, MO 63102
(314) 739-4224 Fax (314) 621-5125
www.fire-investigators.org

International Association of Bomb Technicians and Investigators
P.O. Box 8629
Naples, FL 34101
(941)353-6843 Fax (941)353-6841
www.iabti.org

International Association of Chiefs of Police (IACP)
515 N. Washington St.
Alexandria, VA 22314
(800) 843-4227 Fax (703) 836-4543
www.theiacp.org

International Association of Ethics Trainers
1620 West State Route 434, Suite 164
Longwood, FL 32750
(407) 339-0322 Fax (407) 339-7139
www.ethicstrainers.com

International Association of Financial Crimes Investigators
385 Bel Marin Keys, Suite H
Novato, CA 94949
(415) 884-6600 Fax (415) 884-6605
www.iafci.org

International Association of Fire Chiefs
4025 Fair Ridge Rd., Suite 300
Fairfax, VA 22033
(703) 273-0911 Fax (703) 273-9363
www.iafc.org

International Association of Women Police
P.O. Box 15207
Seattle, WA 98115
(206) 625-4465
www.iawp.org

International Narcotic Enforcement Officers Association Inc.
112 State St., Suite 1200
Albany, NY 12207
(518) 253-2874 Fax (518) 253-3378
www.ineoa.org

National Association of School Resource Officers
2714 S.W. 5th St.
Boynton Beach, FL 33435
(561) 736-1736 Fax (561) 736-1736
www.nasro.org

National Institute of Ethics
1610 West State Route 434, Suite 164
Longwood, FL 32750
(407) 339-0322 Fax (407) 339-7139
www.ethicsinstitute.com

National Institute of Justice Resources
Office of the Director
810 7th St., N.W.
Washington, D.C. 20531
(202) 307-2942 Fax (202) 307-6394
www.ojp.usdoj.gov/nij

National Criminal Justice Reference Service
Office of the Director
P.O. Box 6000
Rockville, MD 20849
(800) 851-3420 Fax (301) 519-5212
http://www.ncjrs.org

Bureau of Justice Assistance Clearinghouse
(800) 688-4252 Fax (410) 953-3848
www.ncjrs.org/task16.htm

Juvenile Justice Clearinghouse
(800) 638-8736 Fax (301) 519-5212
www.fsu.edu/~crimedo/jjclearinghouse

Victims of Crimes Resource Center
(800) 627-6872 Fax (410) 953-3848
www.avp.org/resources.htm

APPENDIX C

LAW ENFORCEMENT-RELATED ASSOCIATIONS

AAA Foundation for Traffic Safety
1440 New York Ave. N.W., Suite 201
Washington, D.C. 20005
(202) 638-5944 Fax (202) 638-5943
www.aaafts.org

Air Force Security Police Association
818 Willow Creek Cir.
San Marcos, TX 78666-5060
(800) 782-7653 Fax (512) 396-7328
http://uts.cc.utexas.edu/~ehre/AFSPA.html

Air Incident Research
P.O. Box 4745
East Lansing, MI 48826
(517) 336-9375 Fax (517) 336-9375
www.airsafety.com

Airborne Law Enforcement Association Inc.
14268 Linda Vista Dr.
Whittier, CA 90602
(213) 485-2011 Fax (213) 485-2073
www.alea.org

American Association of Retired Persons (AARP)
601 E St. N.W.
Washington, D.C. 20049
(202) 434-2222 Fax (202) 434-6466
www.aarp.org

American Association of State Highway and Transportation Officials
444 N. Capitol St. N.W., Suite 249
Washington, D.C. 20001
(202) 434-2222 Fax (202) 434-6466
www.aashto.org

American Correctional Association
4380 Forbes Blvd.
Lanham, MD 20706
(301) 918-1800 Fax (301) 918-1900
www.correctional.com/aca

American Criminal Justice Association
P.O. Box 601047
Sacramento, CA 95860
(916) 484-6553 Fax (916) 488-2227
www.acjalae.org

American Federation of Police and Concerned Citizens
3801 Biscayne Blvd.
Miami, FL 33137
(305) 573-0070 Fax (305) 573-9819
www.aphf.org

American Jail Association
2053 Day Rd., Suite 100
Hagerstown, MD 21740
(301) 790-3930 Fax (301) 790-2941
www.corrections.com/aja

American Planning Association
1776 Massachusetts Ave. N.W., Suite 400
Washington, D.C. 20036
(202) 872-0611 Fax (202) 872-0643
www.planning.org

American Police Hall of Fame and Museum
3801 Biscayne Blvd.
Miami, FL 33137
(305) 573-0070 Fax (305) 573-9819
www.aphf.org

American Police Officers Association
2173 Embassy Dr.
Lancaster, PA 17603
(888) 644-8022
www.apai.org

American Polygraph Association
P.O. Box 8037
Chattanooga, TN 37414
(423) 892-3992 Fax (423) 894-5435
1-800-APA-8037
www.polygraph.org

American Probation and Parole Association
2760 Research Park Dr.
Lexington, KY 40578
(606) 244-3216 Fax (606) 244-8001
www.csom.org

American Psychiatric Association
1400 K St., N.W.
Washington, D.C. 20005
(202) 682-6000 Fax (202) 682-6850
www.psych.org

American Society of Law Enforcement Trainers (ASLET)
P.O. Box 361
Lewes, DE 19958
(302) 645-4080 Fax (302) 645-4084
www.aslet.org

Americans for Effective Law Enforcement Inc.
5519 W. Cumberland Ave., Suite 1008
Chicago, IL 60656
(800) 763-2802 Fax (800) 763-3221
www.aele.org

American Association of Railroads
P.O. Box 11130
Pueblo, CO 81001
(719) 584-0701 Fax (719) 585-1819
www.aar.org

Association of Public Safety Communications Officials International Inc.
2040 S. Ridgeway Ave.
South Daytona, FL 32119
(904) 322-2500 Fax (904) 322-2501
www.apcointl.org

Canada Association of Chiefs of Police (CACP)
130 Albert St., Suite 1710
Ottawa, Canada K1P 5G4
(613) 233-1106 Fax (613) 233-6960
www.cacp.ca

Child Find of America, Inc.
243 Main St.
New Paltz, NY 12561
(845) 255-1848 Fax (845) 255-5706
www.childfindofamerica.org

Child Shield USA
103 W. Spring St.
Titusville, PA 16354
(800) 652- 4453 Fax (814) 827-6977
www.childshieldusa.com

Commission on Accreditation for Law Enforcement Agencies
10306 Eaton Pl., Suite 320
Fairfax, VA 222030
(800) 368-3757 Fax (703) 591-2206
www.calea.org

Concerns of Police Survivors Inc. (COPS)
South Highway 5
Camdenton, MO 65020
(573) 346-4911 Fax (573) 346-1414

Congressional Fire Services Institute
900 Second St. N.E., Suite 303
Washington, D.C. 20002
(202) 371-1277 Fax (202) 682-3473
www.cfsi.org

Council of International Investigators
2150 N. 107th, Suite 205
Seattle, WA 98133-9009
www.cii2.org

Crime Stoppers International, Inc.
P.O. Box 614
Arlington, TX 76004-0614

D-A-R-E America
P.O. Box 512090
Los Angeles, CA 90051
(310) 215-0575 Fax (310) 215-0180
www.dare-America.com

Eastern Armed Robbery Conference, Ltd.
P.O. Box 5772
Wilmington, DE 19808
(516) 852-6271 Fax (516) 852-6478
www.earc.org

Federal Law Enforcement Officers Association
P.O. Box 508
East Northport, NY 11731
(717) 938-2300 Fax (717) 932-2262
www.fleoa.org

Federal Pretrial Services Agency
500 Pearl St., Room 550
New York, NY 10007
(212) 805-0015 Fax (212) 805-4172

Hostage Negotiators of America
2072 Edinburgh Dr.
Montgomery, AL 36116
(334) 244-7411

Institute of Police Technology and Management
University of North Florida
12000 Alumni Dr.
Jacksonville, FL 32224-2678
(904) 620-4786 Fax (904) 620-2453
www.iptm.org

International Association of Ethics Trainers
1620 West State Route 434, Suite 164
Longwood, FL 32750
(407) 339-0322 Fax (407) 339-7139
www.ethicstrainers.com

International Association for Identification
2535 Pilot Knob Rd., Suite 117
Mendota Heights, MN 55120
(651) 681-8566 Fax (651) 681-8443
www.theiai.org

International Association for Property and Evidence
903 N. San Fernando Blvd., Suite 4
Burbank, CA 91504-4327
(800) 449-4273 Fax (818) 846-4543
www.iape.org

International Association of Bomb Technicians and Investigators
P.O. Box 8629
Naples, FL 34101
(941) 353-6843 Fax (941) 353-6841
www.iabti.org

International Association of Campus Law Enforcement Administrators (IACLEA)
342 N. Main St.
West Hartford, CT 06117
(860) 586-7517 Fax (860) 586-7550
www.iaclea.org

International Association of Chiefs of Police (IACP)
515 N. Washington St.
Alexandria, VA 22314
(800) 843-4227 Fax (703) 836-4543
www.theiacp.org

International Association of Correctional Officers
3900 Industrial Ave.
Lincoln, NE 68504
(402) 464-0602 Fax (402) 464-5931

International Association of Financial Crimes Investigators
385 Bel Marin Keys Blvd., Suite H
Novato, CA 94949
(415) 884-6600 Fax (415) 884-6605
www.iatci.org

International Association of Fire Chiefs
4025 Fair Ridge Rd., Suite 300
Fairfax, VA 22033
(703) 273-0911 Fax (703) 273-9363
www.iafc.org

International Association of Law Enforcement Planners
1300 Executive Center Dr., Suite 450
Tallahassee, FL 32301-5025
(850) 878-7254 Fax (850) 656-7944
www.ialep.org

International Association of Personal Protection Agents
458 W. Kenwood
Brighton, TN 38011-6294
(901) 837-1915 Fax (901) 837-4949
www.tiac.net/users/jmking

International Association of Women Police
P.O. Box 15207
Seattle, WA 98115
(206) 625-4465
www.iawp.org

International City/County Management Association (ICMA)
777 N. Capitol St. N.E., Suite 500
Washington, D.C. 20002
(202) 289-4262 Fax (202) 962-3500
www.icma.org

International Conference of Police Chaplains
P.O. Box 5590
Destin, FL 32540
(850) 654-9736 Fax (850) 654-9742
http://members.tripod.com/~FSG3/icpc.htm

International Fire Marshals Association
One Batterymarch Park
Quincy, MA 02269
(617) 984-7424 Fax (617) 984-7056

International Foundation for Art Research
500 5th Ave., Suite 1234
New York, NY 10101
(212) 391-6234 Fax (212) 391-8794
www.ifar.com

International Foundation for Protection Officers
3106 Tami Annie Trail
Naples, FL 34103
(941) 430-0534 Fax (941) 430-5333
www.ifpo.com

International Juvenile Officers Association Inc.
P.O. Box 56
Easton, CT 06612
(203) 377-4424

International Law Enforcement Stress Association
5485 David Blvd.
Port Charlotte, FL 33981
(813) 697-8863

International Narcotic Enforcement Officers Association Inc.
112 State St., Suite 1200
Albany, NY 12207
(518) 253-2874 Fax (518) 253-3378
www.ineoa.org

International Police Association–U.S. Section
P.O. Box 43-1822
Miami, FL 33243
(305) 253-2874 Fax (305) 253-3568
www.ipausa.org

International Police Mountain Bike Association
28 E. Ostend St.
Baltimore, MD 21230
(410) 685-2220 Fax (410) 685-2240
www.ipmba.org

International Prisoners Aid Association
Dept. of Sociology, University of Louisville
Louisville, KY 40292
(502) 241-7831 Fax (502) 241-7831

International Union of Police Association AFL/CIO
1421 Prince St., Suite 330
Alexandria, VA 2314
(703) 549-7434 Fax (703) 549-9048

Jewelers Security Alliance
6 E. 45th St.
New York, NY 10017
(800) 537-0067 Fax (212) 808-9168
http://jsa.polygon.net

Law Enforcement Alliance of America
7700 Leesburg Pike, Suite 421
Falls Church, VA 22043
(703) 847-2677 Fax (703) 556-6485
www.leaa.org

Law Enforcement and Emergency Services Video Association
P.O. Box 126167
Fort Worth, TX 76126
(817) 249-4002 Fax (817) 249-4002

Law Enforcement Memorial Association Inc.
P.O. Box 72835
Roselle, IL 60172
(847) 795-1547 Fax (847) 795-2469
www. w8ca.com/lema/

Military Police Regimental Association
P.O. Box 5278
Anniston, AL 36205
(256) 848-5014 Fax (256) 848-6691

Missing Children Society of Canada
3501 23rd St., N.E. Suite 219
Calgary, Canada T2E 6V8
(403) 291-0705 Fax (403) 291-9728
www.mcsc.ca/

Narcotic Enforcement Officers Association
29 N. Plains Highway Phoenix Park, Suite 10
Wallingford, CT 06492
(203) 269-8940 Fax (203) 284-9103
www.neoa.org

National Association Against Gang and Domestic Violence
P.O. Box 775186
Saint Louis, MO 63177
(314) 631-3723

National Association of Chiefs of Police
1000 Connecticut Ave. N.W., Suite 9
Washington, D.C. 20036
(202) 293-9088 Fax (305) 573-9819
http://w3.bluegrass.net/~n2try/nacop.htm

National Association of Counties
440 First St. N.W., 8th Floor
Washington, D.C. 20001
(202) 393-6226 Fax (202) 393-2630
www.naco.org

National Association of Drug Court Professionals
901 N. Pitt St., Suite 370
Alexandria, VA 22314
(703) 706-0576 Fax (703) 706-0577
www.drugcourt.org/home.cfm

National Association of Field Training Officers (NAFTO)
Sage Valley Rd.
Longmont, CO 80503
(303) 442-0482 Fax (303) 546-6791
www.nafto.org

National Association of Medical Examiners (N.A.M.E.)
1402 S. Grand Blvd.
St. Louis, MO 3104
(314) 577-8298
www.thename.org

National Association of Police Athletic Leagues
618 N. U.S. Highway 1, Suite 201
N. Palm Beach, FL 33408
(561) 844-1823 Fax (561) 863-6120
www.reeusda.gov/pavnet/cj/cjnatass.htm

National Association of Police Organizations (NAPO)
750 First St., Suite 920
Washington, D.C. 20002
(202) 842-4420 Fax (202) 842-4396
www.napo.org

National Association of School Resource Officers (NASRO)
2714 S.W. 5th St.
Boynton Beach, FL 33435
(561) 736-1736 Fax (561) 736-1736
www.nasro.org

National Association of Town Watch (NATW)
One Wynnewood Rd., Suite 102
Wynnewood, PA 19096
(610) 649-7055 Fax (610) 649-5456
www.nationaltownwatch.org

National Center for Missing and Exploited Children
2101 Wilson Blvd., Suite 550
Arlington, VA 22201
(703) 235-3900 Fax (703) 235-4067
www.missingkids.com

National Child Safety Council
4065 Page Ave.
P.O. Box 1368
Jackson, MI 49204
(517) 764-6070 Fax (517) 764-4140
(800) 222-1464

National Constables Association
16 Stonybrook Dr.
Levittown, PA 19055
(215) 547-6400 Fax (215) 943-0907
www.angelfire.com/la/nationalconstable

National Council on Crime and Delinquency (NCCD)
685 Market St., Suite 620
San Francisco, CA 94105
(415) 896-6223 Fax (415) 896-5109
www.nccd-crc.org

National Crime Prevention Council
1700 K St., 2nd Floor
Washington, D.C. 20006
(202) 466-6272 Fax (202) 296-1356
www.ncpc.org

National Crime and Punishment Learning Center
623 Sarazen Dr.
Gulfport, MS 39507
(228) 896-5280 Fax (228) 896-8696
http://crimeandpunishment.net

National Criminal Justice Association
444 N. Capitol St. N.W., Suite 618
Washington, D.C. 20001
(202) 624-1440 Fax (202) 508-3859
www.sso.org/ncja/default.htm

National District Attorneys Association
99 Canal Center Plaza, Suite 510
Alexandria, VA 22314
(703) 549-9222 Fax (703) 836-3195
www.ndaa.org

National Family Legal Foundation (NFLF)
11000 N. Scottsdale Rd., Suite 144
Scottsdale, AZ 85254
(602) 922-9731 Fax (602) 922-7240
www.nflf.com

National Fire Protection Association
One Batterymarch Park
Quincy, MA 02269
(617) 770-3000 Fax (617) 770-0700
www.nfpa.org

National Fraternal Order of Police
1410 Donelson Pike, Suite A17
Nashville, TN 37217
(615) 339-0900 Fax (615) 339-0400
www.grandlodgefop.org

National Institute of Ethics
1610 West State Route 434, Suite 164
Longwood, FL 32750
(407) 339-0322 Fax (407) 339-7139
www.ethicsinstitute

National Institute on Economic Crime
P.O. Box 7186
Fairfax, VA 22039
(703) 250-8706 Fax (703) 860-8449

National Insurance Crime Bureau (NICB)
10330 S. Roberts Rd.
Palos Hills, IL 60465
(708) 430-2430 Fax (708) 430-2446
www.nicb.org

National Law Enforcement Council
888 Sixteenth St. N.W., Suite 700
Washington, D.C. 20006
(202) 835-8020 Fax (202) 331-4291

National Law Enforcement Officers Memorial Fund Inc.
605 E St., N.W.
Washington, D.C. 20004
(202) 737-3400 Fax (202) 737-3405
www.nleomf.com

National Law Enforcement Research Center
P.O. Box 70966
Sunnydale, CA 94086
(408) 245-2037 Fax (408) 245-2037

National League of Cities (NLC)
1301 Pennsylvania Ave. N.W., Suite 550
Washington, D.C. 20004
(202) 626-3000 Fax (202) 626-3043
www.nlc.org

National Legal Aid and Defender Association
1625 K St. N.W., Suite 800
Washington, D.C. 20006
(202) 452-0620 Fax (202) 872-1031
www.nlada.org

National Organization for Victim Assistance
1757 Park Rd. N.W.
Washington, D.C.
(202) 232-6682 Fax (202) 462-2255
www.try-nova.org

National Organization of Black Law Enforcement Executives
1757 Park Rd. N.W.
Washington, D.C. 20010
(703) 658-1529 Fax (703) 658-9479
www.noblenatl.org

National Police Institute
Central Missouri State University
200 Main St.
Warrensburg, MO 22312
(660) 543-4091 Fax (660) 543-83306

National Recreation and Park Association
DuPage County Forest Preserve District
Glen Ellyn, IL 60138
(630) 933-7239 Fax (630) 790-1071

National Reserve Law Officers Association
P.O. Box 6505
San Antonio, TX 78209
(210) 820-0478 Fax (210) 804-2463

National Rifle Association
11250 Waples Mill Rd.
Fairfax, VA 22030
(703) 267-1000 Fax (703) 267-3989
www.nra.org/

National Safety Council
1025 Connecticut Ave. N.W., Suite 1200
Washington, D.C. 20036
(202) 974-2480 Fax (202) 293-0032
www.nsc.org

National Sheriffs Association
1450 Duke St.
Alexandria, VA 22314
(703) 836-7827 Fax (703) 683-6541
www.sheriffs.org

National Tactical Officers Association (NTOA)
P.O. Box 529
Doylestown, PA 18901
(800) 279-9127 Fax (215) 230-7552
www.ntoa.org

National Technical Investigators Association
6933 N. 26th St.
Falls Church, VA 22046
(703) 237-9388 Fax (703) 241-0353
www.natia.org

National Traffic Law Center
99 Canal Center Plaza, Suite 510
Alexandria, VA 22314
(703) 549-4253 Fax (703) 836-3195

National Troopers Coalition
Andrea Lane
La Plata, MD 20646
(410) 653-3885 Fax (410) 653-0929
www.nat-trooperscoalition.com

National United Law Enforcement Officers Association Inc.
265 E. McLemore Ave.
Memphis, TN 38106
(901) 774-1118 Fax (901) 774-1139

National White Collar Crime Center
7401 Beausant Springs Dr.
Richmond, VA 23225
(804) 323-3563 Fax (804) 323-3566
www.nw3c.org

National Wildlife Federation
8925 Leesburg Pike
Vienna, VA 22184
(703) 790-4000 Fax (703) 790-4330
www.hwf.org

National Youth Gang Center (NYGC)
2894 Remington Green Circle
Tallahassee, FL 32308
(850) 385-0600 Fax (850) 385-5356
www.iir.com/nygc

Nine Lives Associates Executive Protection Institute
Rural Route 1, Box 332
Bluemont, VA 26135
(540) 955-1128 Fax (540) 955-0255

Office of International Criminal Justice (OICJ)
1033 W. Van Burden St.
Chicago, IL 60607
(312) 996-9595 Fax (312) 312-0458
www.acsp.uic.edu/

Office of Law Enforcement Standards (OLES)
National Institute of Standards and Technology Building, Room 225
Gaithersburg, MD 20899
(800) 975-2757 Fax (301) 948-0978
www.oles.org

Operation Lifesaver
1420 King St., Suite 401
Alexandria, VA 22314
(800) 537-6224 Fax (703) 519-8267
www.oli.org

Operation Lookout National Center for Missing Youth
6320 Evergreen Way, Suite 201
Everett, WA 98203
(800) 782-7335 Fax (425) 438-4111
www.operationlookout.org

Organized Crime Task Force
143 Grand St.
White Plains, NY 10601
(914) 422-8780 Fax (914) 422-8795

Police Chiefs Spouses—Worldwide
1521 Sixth Ave. East
Menomonie, WI 54751
(715) 235-9749
www.gtesupersite.com/pcsw/

Police Communication Center
215 Church Ave. S.W.
Roanoke, VA 24011
(540) 853-2411 Fax (540) 853-1599

Police Executive Research Forum (PERF)
1120 Connecticut Ave. N.W., Suite 930
Washington, D.C. 20036
(202) 466-7820 Fax (202) 466-7826
www.inca.net/perf

Police and Fireman's Insurance Association
101 E. 116th St.
Carmel, IN 46032
(317) 581-1913 Fax (317) 571-5946

Police Marksman Association
6000 E. Shirley Lane
Montgomery, AL 36117
(334) 271-2010 Fax (334) 279-9267
www.policemarksman.com

Pretrial Services Resource Center
1325 G St. N.W., Suite 770
Washington, D.C. 20005
(202) 638-3080 Fax (202) 347-0493
www.pretrial.org

Reserve Law Officers Association of America (RESLAW)
San Antonio, TX 78217
(210) 653-5754 Fax (210) 653-9655
www.reslaw.com

Retired and Disabled Police of America
1900 S. Harbor City Blvd., Suite 328
Melbourne, FL 32901
(800) 395-7376 Fax (407) 779-8046

The Police Supervisors Group
1401 Johnson Ferry Rd., Suite 328-F42
Marietta, GA 30062
(770) 321-5018 Fax (770) 321-5019
www.policesupervisors.org

Transportation Research Board
2101 Constitution Ave. N.W.
Washington, D.C. 20418
(202) 334-2936 Fax (202) 334-2003
www.nas.edu.trb

U.S. Conference of Mayors
1620 I St. N.W.
Washington, D.C. 20006
(202) 293-7330 Fax (202) 293-2352
www.USmayors.org

United States Fire Administration (USFA)
National Emergency Training Center
16825 S. Seton
Emmitsburg, MD 21727
(301) 447-1200 Fax (301) 447-1102
wwwusfa.fema.gov

APPENDIX D

STATE TRAINING COUNCILS AND POLICE STANDARDS COMMISSIONS

Contact your state training council or police standards commission to learn of upcoming training opportunities or for questions about the hiring or training requirements in your state.

R. Alan Benefield, Chief Director
Alabama Peace Officer Standards and Training Board
P.O. Box 300075
Montgomery, AL 35130-0075
(334) 242-4045 Fax (334) 242-4633
www.apostc.state.al.us

Irl T. Stambaugh Executive Director
Alaska Police Standards Council
P.O. Box 111200
Juneau, AK 99811-1200
(907) 465-4378 Fax (907) 465-3263
www.dps.state.ak.us/aps

Rod Covey, Executive Director
Arizona Peace Officer Standards and Training Board
2643 East University
P.O. Box 6638
Phoenix, AZ 85005
(602) 223-2514 Fax (602) 244-0477
www.azpost.state.az.us

Terry Bolton, Director
Arkansas Commission on Law Enforcement Standards and Training
P.O. Box 3106
East Camden, AR 71711
(870) 574-1810 Fax (870) 574-2706
www.law/enforcement.org/CLEST

Kenneth O'Brien, Executive Director
California Commission on Peace Officer Standards and Training
1601 Alahambra Blvd.
Sacramento, CA 95816-7053
(916) 227-2802 Fax (916) 227-2801
www.POST.ca.gov

John Kammerzell, Director
Colorado Peace Officer Standards and Training Board
1525 Sherman St., 6th Floor
Denver, CO 80203
(303) 866-5692 Fax (303) 866-5671
www.ago.state.co.us/post/pst

T. William Knapp, Executive Director
Connecticut Police Officer Standards and Training Council
285 Preston Ave.
Meridan, CT 06450
(203) 238-6505 Fax (203) 238-6503
www.post.state.ct.us

Col. James Ford, Jr. (Ret.), Chairman
Delaware Council on Police Training
1453 DuPont Highway
P.O. Box 430
Dover, DE 19903
(302) 739-5903 Fax (302) 739-5945

A. Leon Lowry, II, Director
Florida Dept. of Law Enforcement
P.O. Box 1489
Tallahassee, FL 32302-1489
(850) 410-8600 Fax (850) 410-8606

Steve Black, Executive Director
Georgia Peace Officer Standards and Training Council
2175 Northlake Parkway, Suite 144
Tucker, GA 30084
(770) 414-3313 Fax (770) 414-3332
www.gapost.org

Eugene Vemura, Assistant Chief
Honolulu Police Department
801 S. Beretania St.
Honolulu, HI 96813
(808) 677-1474 Fax (808) 677-7394
www.honolulupd.org

Michael N. Becar, Executive Director
Idaho Peace Officer Standards and Training
P.O. Box 700
Meridian, ID 83680-0700
(208) 884-7250 Fax (208) 884-7295
www.idaho.post.org

Thomas J. Jurkanin, Executive Director
Illinois Law Enforcement Training and Standards Board
600 S. 2nd. St.
Springfield, IL 62704-2542
(217) 782-4540 Fax (217) 524-5711
www.ptb.state.il.us

Charles C. Burch, Executive Director
Indiana Law Enforcement Training Board
P.O. Box 313
Plainfield, IN 46168-0313
(317) 839-5191 ext. 212 Fax (317) 839-9741
www.state.in.us/iu

Gene W. Shepard, Director
Iowa Law Enforcement Academy
P.O. Box 130
Johnston, IA 50131-0130
(515) 242-5357 Fax (515) 242-5471
www.state.ia.us/ilea

Ed H. Pavey, Director
Kansas Law Enforcement Training Center
P.O. Box 647
Hutchinson, KS 67504
(316) 662-3378 Fax (316) 662-4720
www.kletc.org

John Bizzack, Commissioner
Kentucky Dept. of Criminal Justice Training
Kit Carson Dr., Funderbach Bldg., EKU
Richmond, KY 40475
(606) 622-1328 Fax (606) 622-2740
www.docjt.jus.state.ky.us

Michael Ranatza, Director
Louisiana Commission on Law Enforcement
1885 Wooddale Blvd., Room 208
Baton Rouge, LA 70806
(225) 925-4942 Fax (225) 925-1998
www.cole.State.la.us

Steven R. Giorgetti, Director
Maine Criminal Justice Academy
93 Silver St.
Waterville, ME 04901
(207) 877-8000 Fax (207) 877-8027
www.state.maine.us

Donald Hopkins, Executive Director
Maryland Police and Correctional Training Commission
3085 Hemwood Rd.
Woodstock, MD 21163
(410) 750-6500 Fax (410) 203-1010
www.dpscs.state.md.us/pct

Massachusetts Criminal Justice Training Council
411 Waverly Oaks Rd. Suite 250
Waltham, MA 02154
(617) 727-7827 Fax (617) 642-6898
www.state.ma.us/cjtc

Raymond Beach, Jr., Executive Director
Michigan Commission on Law Enforcement Standards
7426 N. Canal Rd.
Lansing, MI 48913
(517) 322-1946 Fax (517) 322-6439
www.coles-online.org

John T. Laux, Executive Director
Minnesota Board of Police Officers Standards and Training
1600 University Ave., Suite 200
St. Paul, MN 55104
(612) 543-3060 Fax (612) 643-3072

Chris Egbert, Program Administrator
Missouri Peace Officer Standards and Training
P.O. Box 749
Jefferson City, MO 65102
(573) 751-4819 Fax (573) 517-5399
www.dps.state.mo/us

James B. Walker, Director
Mississippi Office on Law Enforcement Standards and Training
401 Northwest St.
P.O. Box 23039
Jackson, MS 39204
(601) 359-7880 Fax (601) 359-7832

Jim Oberhofer, Executive Director
Montanta Peace Officer Standards and Training Conncil
303 N. Roberts
Helena, MT 59620
(406) 444-3604 (406) 444-4722
www.adlest.org/montana

Steve Lamken, Director
Nebraska Law Enforcement Training Center
3600 North Academy Rd.
Grand Island, N.E. 68801
(308) 385-6030 Fax (308) 385-6032
www.nletc.state.ne.us

Richard Clark, Executive Director
Nevada Commission Peace Officer's Standards and Training
3476 Executive Pointeway
Carson City, NV 89706
(775) 684-7678 Fax (775) 687-4911
www.leg.state.NV.us/nac

Earl M. Sweeney, Director
New Hampshire Police Standards and Training Council
17 Fan Rd.
Concord, NH 03301
(603) 271-2133 Fax (603) 271-1785
www.justiceworks.unh.edu

Wayne Fisher, Deputy Director
New Jersey Police Training Commission
P.O. Box 085
Trenton, NJ 07039
(609) 984-0960 Fax (609) 984-4473
www.state.nj.us/lps/dcj

Darrel G. Hart, Director
New Mexico Law Enforcement Academy Board
4491 Cerrillos Rd.
Santa Fe, NM 87505
(505) 827-9255 Fax (505) 827-3449
www.dps.nm.org/training

Jerry Burrell, Deputy Commissioner
New York Office of Public Safety—Division of Criminal Justice Services
4 Tower Place
Albany, NY 12203
(518) 457-6101 Fax (518) 457-0145
www.criminaljust.state.ny.us

David D. Cashwell, Director
North Carolina Criminal Justice Standards Division
P.O. Drawer 149
(919) 716-6470 Fax (919) 716-6752
Raleigh, NC 27602
www.jus.state.nc.us/justice

Mark Gilbertson, Executive Secretary
North Dakota Peace Officer Standards and Training Law Enforcement Academy
600 East Blvd., Dept. 504
Bismarck, ND 58505
(701) 328-9968 Fax (701) 328-9988
www.iadlest.org/ndakota

Vernon Chenevey, Executive Director
Ohio Peace Officer Training Commission
1650 State Route 56 N.W.
P.O. Box 309
London, OH 43140
(614) 466-7771 Fax (614) 728-5150
www.ag.state.oh.us

Jeanie Nelson, Ph.D., Director
Oklahoma Council on Law Enforcement Education and Training
P.O Box 11476
Oklahoma City, OK 73136
(405) 425-2750 Fax (405) 425-2773
www.dps.state.ok.us/clete

Dianne Middle, Director
Dept. of Public Safety Standards and Training
550 N. Monmouth Ave.
Monmouth, OR 97361
(503) 378-2100 Fax (503) 378-3330
www.oregonvof.net/dpsst

Richard Mooney, Executive Director
Pennsylvania Municipal Police Officer's Education and Training Commission
75 E. Derry Rd.
Hershey, PA 17033
(717) 533-5987 Fax (717) 787-1650
www.mpoetc.org

Steven D. Weaver, Director
Rhode Island Municipal Police Training Academy, Community College of Rhode Island—Flanagan Campus
Lincoln, RI 02865
(401) 222-3755 Fax (401) 726-5720

William Gibson, Deputy Director
South Carolina Criminal Justice Academy
5400 Broad River Rd.
Columbia, SC 29210
(803) 896-7777 Fax (803) 896-8347
www.iadlest.org/scaa

Kevin Thorn, Director
South Dakota Law Enforcement Standards and Training Commission
E. Hwy. 34, 500 East Capital
Pierre, SD 57501
(605) 773-3584 Fax (605) 773-4629

Mark Bracy, Executive Secretary
Tennessee Peace Officer Standards and Training Commission
3025 Lebanon Rd.
Nashville, TN 37214
(615) 741-3361 Fax (615) 532-0502

Jim Dozier, Executive Director
Texas Commission on Law Enforcement Officer Standards and Education
6300 U.S. Hwy. 290 E., Suite 200
Austin, TX 78723
(512) 936-7700 Fax (512) 406-3614
www.tcjeose.state.tx.us

Sidney Groll, Executive Director
Utah Peace Officer Standards and Training
4525 S. 2700 West
Salt Lake City, UT 84119
(801) 965-4731 Fax (801) 965-4619
www.ps.ex.state.ut.us/pos

Gary L. Bullard, Executive Director
Vermont Criminal Justice Training Council
Rural Rd. 2, Box 2160
Pittsford, VT 05487
(802) 483-6225 Fax (802) 483-2343
www.vcjt.state.vt.us

George Gotschalk, Deputy Director
Virginia Dept. of Criminal Justice Services
805 E. Broad St.
Richmond, VA 23219
(804) 786-4000 Fax (804) 786-0588
www.dcjs.state.va.us

Michael Parsons, Ph.D., Executive Director
Washington State Criminal Justice Training Commission
19010 1st Ave. S.
Seattle, WA 98148
(206) 439-3740 Fax (206) 439-3752
www.wa.gov/cjt

Don Davidson, Training Coordinator
West Virginia Criminal Justice Services
1204 Kanawha Blvd. E.
Charleston, West VA 25301
(304) 558-8814 ext. 214 Fax (304) 558-0391
www.wvdjs.com

Dennis Hanson, Director
Wisconsin Dept. of Justice, Training and Standards Bureau
P.O. Box 7070
Madison, WI 53707
(608) 266-8800 Fax (608) 266-7869

Donald Pierson, Executive Director
Wyoming Peace Officer Standards and Training
1710 Pacific Ave.
Cheyenne, WY 82002
(307) 777-6619 Fax (307) 638-9706
www.iadlest.org/wyoming

APPENDIX E

County	Name of agency	Minimum educational level required 4-year degree	2-year degree	Some college*	H.S. diploma
ALABAMA					
Etowah	Gadsden Police	-	-	-	X
Houston	Dothan Police	-	-	-	X
Jefferson	Jefferson County Sheriff	-	-	-	X
Jefferson	Bessemer Police	-	-	-	X
Jefferson	Birmingham Police	-	-	-	X
Jefferson	Hoover Police	-	-	-	X
Madison	Huntsville Police	-	-	X	-
Mobile	Mobile County Sheriff	-	-	-	X
Mobile	Mobile Police	-	-	-	X
Montgomery	Montgomery County Sheriff	-	-	-	X
Montgomery	Montgomery Police	-	-	-	X
Morgan	Decatur Police	-	-	-	X
Tuscaloosa	Tuscaloosa Police	-	-	-	X
ALASKA					
Anchorage	Anchorage Police	-	-	-	X
ARIZONA					
Maricopa	Maricopa County Sheriff	-	-	-	X
Maricopa	Chandler Police	-	-	-	X
Maricopa	Glendale Police	-	-	-	X
Maricopa	Mesa Police	-	-	-	X

County	Name of agency	Minimum educational level required			
		4–year degree	2–year degree	Some college*	H.S. diploma
Maricopa	Phoenix Police	-	-	-	X
Maricopa	Scottsdale Police	-	-	X	-
Maricopa	Tempe Police	-	X	-	-
Pima	Pima County Sheriff	-	-	-	X
Pima	Tucson Police	-	-	-	X
Pinal	Pinal County Sheriff	-	-	-	X
Yuma	Yuma Police	-	-	-	X
ARKANSAS					
Jefferson	Pine Bluff Police	-	-	-	X
Pulaski	Pulaski County Sheriff	-	-	-	X
Pulaski	Little Rock Police	-	-	-	X
Pulaski	North Little Rock Police	-	-	-	X
Sebastian	Fort Smith Police	-	-	-	X
CALIFORNIA					
Alameda	Alameda County Sheriff	-	-	-	X
Alameda	Alameda Police	-	-	-	X
Alameda	Berkeley Police	-	X	-	-
Alameda	Fremont Police	-	-	X	-
Alameda	Hayward Police	-	-	-	X
Alameda	Oakland Police	-	-	-	X
Contra Costa	Contra Costa County Sheriff	-	-	-	X
Contra Costa	Concord Police	-	-	X	-

County	Name of agency	Minimum educational level required 4-year degree	2-year degree	Some college*	H.S. diploma
Contra Costa	Richmond Police	-	-	X	-
El Dorado	El Dorado County Sheriff	-	-	-	X
Fresno	Fresno County Sheriff	-	-	X	-
Fresno	Fresno Police	-	-	-	X
Kern	Kern County Sheriff	-	-	-	X
Kern	Bakersfield Police	-	-	-	X
Los Angeles	Los Angeles County Sheriff	-	-	-	X
Los Angeles	Beverly Hills Police	-	-	X	-
Los Angeles	Burbank Police	-	-	-	X
Los Angeles	Compton Police	-	-	-	X
Los Angeles	Culver City Police	-	-	-	X
Los Angeles	Downey Police	-	-	-	X
Los Angeles	El Monte Police	-	-	-	X
Los Angeles	Glendale Police	-	X	-	-
Los Angeles	Inglewood Police	-	-	-	X
Los Angeles	Long Beach Police	-	-	-	X
Los Angeles	Los Angeles Police	-	-	-	X
Los Angeles	Pasadena Police	-	-	-	X
Los Angeles	Pomona Police	-	-	-	X
Los Angeles	Santa Monica Police	-	-	-	X
Los Angeles	Torrance Police	-	-	-	X
Los Angeles	West Covina Police	-	-	X	-
Los Angeles	Whittier Police	-	-	-	X

County	Name of agency	Minimum educational level required			
		4–year degree	2–year degree	Some college*	H.S. diploma
Marin	Marin County Sheriff	-	-	-	X
Monterey	Monterey County Sheriff	-	-	-	X
Monterey	Salinas Police	-	-	-	X
Orange	Orange County Sheriff-Coroner	-	-	-	X
Orange	Anaheim Police	-	-	-	X
Orange	Costa Mesa Police	-	-	-	X
Orange	Fullerton Police	-	-	-	X
Orange	Garden Grove Police	-	-	-	X
Orange	Huntington Beach Police	-	-	-	X
Orange	Irvine Police	-	X	-	-
Orange	Newport Beach Police	-	-	-	X
Orange	Orange Police	-	-	-	X
Orange	Santa Ana Police	-	-	-	X
Placer	Placer County Sheriff	-	-	-	X
Riverside	Riverside County Sheriff	-	-	-	X
Riverside	Corona Police	-	-	-	X
Riverside	Riverside Police	-	-	-	X
Sacramento	Sacramento County Sheriff	-	-	-	X
Sacramento	Sacramento Police	-	-	X	-
San Bernardino	San Bernardino County Sheriff	-	-	-	X

County	Name of agency	Minimum educational level required			
		4–year degree	2–year degree	Some college*	H.S. diploma
San Bernardino	Fontana Police	-	-	X	-
San Bernardino	Ontario Police	-	-	-	X
San Bernardino	Rialto Police	-	-	-	X
San Bernardino	San Bernardino Police	-	-	-	X
San Diego	San Diego County Sheriff	-	-	-	X
San Diego	Chula Vista Police	-	-	-	X
San Diego	El Cajon Police	-	-	-	X
San Diego	Escondido Police	-	-	-	X
San Diego	Oceanside Police	-	X	-	-
San Diego	San Diego Police	-	-	-	X
San Francisco	San Francisco Police	-	-	-	X
San Joaquin	San Joaquin County Sheriff	-	-	-	X
San Joaquin	Stockton Police	-	-	-	X
San Luis Obispo	San Luis Obispo County Sheriff	-	-	-	X
San Mateo	San Mateo County Sheriff	-	-	-	X
San Mateo	Daly City Police	-	-	-	X
San Mateo	San Mateo Police	-	-	X	-
Santa Barbara	Santa Barbara County Sheriff	-	-	-	X
Santa Barbara	Santa Barbara Police	-	-	-	X
Santa Clara	Santa Clara County Sheriff	-	-	X	-
Santa Clara	San Jose Police	-	-	X	-

County	Name of agency	Minimum educational level required			
		4–year degree	2–year degree	Some college*	H.S. diploma
Santa Clara	Santa Clara Police	-	-	X	-
Santa Clara	Sunnyvale Police	-	-	X	-
Santa Cruz	Santa Cruz County Sheriff	-	-	-	X
Shasta	Shasta County Sheriff	-	-	-	X
Solano	Vallejo Police	-	-	-	X
Sonoma	Sonoma County Sheriff-Coroner	-	-	-	X
Sonoma	Santa Rosa Police	-	-	-	X
Stanislaus	Stanislaus County Sheriff	-	-	-	X
Stanislaus	Modesto Police	-	-	X	-
Tulare	Tulare County Sheriff	-	-	-	X
Ventura	Ventura County Sheriff	-	-	-	X
Ventura	Simi Valley Police	-	-	-	X
Ventura	Ventura Police	-	-	-	X
COLORADO					
Adams	Adams County Sheriff	-	-	-	X
Adams	Westminster Police	-	-	-	X
Arapahoe	Aurora Police	-	-	X	-
Boulder	Boulder County Sheriff	-	-	-	X
Boulder	Boulder Police	-	X	-	-
Denver	Denver Police	-	-	-	X
Douglas	Douglas County Sheriff	-	-	-	X

County	Name of agency	Minimum educational level required			
		4-year degree	2-year degree	Some college*	H.S. diploma
El Paso	El Paso County Sheriff	-	-	-	X
El Paso	Colorado Springs Police	-	-	X	-
Jefferson	Jefferson County Sheriff	-	-	-	X
Jefferson	Arvada Police	X	-	-	-
Jefferson	Lakewood Police	X	-	-	-
Larimer	Larimer County Sheriff	-	-	-	X
Larimer	Fort Collins Police	-	-	X	-
Pueblo	Pueblo Police	-	-	-	X
CONNECTICUT					
Fairfield	Bridgeport Police	-	-	-	X
Fairfield	Danbury Police	-	-	-	X
Fairfield	Greenwich Police	-	-	-	X
Fairfield	Norwalk Police	-	-	-	X
Fairfield	Stamford Police	-	-	X	-
Hartford	Bristol Police	-	-	-	X
Hartford	East Hartford Police	-	-	-	X
Hartford	Hartford Police	-	-	-	X
Hartford	Manchester Police	-	-	-	X
Hartford	New Britain Police	-	-	-	X
New Haven	Meriden Police	-	-	-	X
New Haven	Milford Police	-	-	-	X
New Haven	New Haven Police	-	-	-	X
New Haven	Waterbury Police	-	-	-	X

County	Name of agency	Minimum educational level required 4–year degree	2–year degree	Some college*	H.S. diploma
New Haven	West Haven Police	-	-	-	X
DELAWARE					
New Castle	New Castle County Police	-	-	X	-
New Castle	Wilmington Police	-	-	-	X
DISTRICT OF COLUMBIA					
Washington, DC	Washington Metropolitan Police	-	-	-	X
FLORIDA					
Alachua	Alachua County Sheriff	-	-	-	X
Alachua	Gainesville Police	-	X	-	-
Bay	Bay County Sheriff	-	-	X	-
Brevard	Brevard County Sheriff	-	-	-	X
Brevard	Melbourne Police	-	-	-	X
Brevard	Palm Bay Police	-	-	-	X
Broward	Broward County Sheriff	-	-	-	X
Broward	Coral Springs Police	-	-	X	-
Broward	Davie Police	-	-	-	X
Broward	Fort Lauderdale Police	-	-	-	X
Broward	Hollywood Police	-	-	-	X
Broward	Margate Police	-	-	-	X
Broward	Miramar Police	-	-	-	X
Broward	Pembroke Pines Police	-	-	-	X
Broward	Plantation Police	-	-	-	X
Broward	Pompano Beach Police	-	-	-	X

County	Name of agency	Minimum educational level required			
		4–year degree	2–year degree	Some college*	H.S. diploma
Broward	Sunrise Police	-	-	-	X
Charlotte	Charlotte County Sheriff	-	-	-	X
Citrus	Citrus County Sheriff	-	-	-	X
Clay	Clay County Sheriff	-	-	-	X
Collier	Collier County Sheriff	-	-	-	X
Dade	Metro-Dade Police	-	-	-	X
Dade	Coral Gables Police	-	-	X	-
Dade	Hialeah Police	-	-	-	X
Dade	Miami Police	-	-	-	X
Dade	Miami Beach Police	-	-	-	X
Duval	Jacksonville Sheriff	-	X	-	-
Escambia	Escambia County Sheriff	-	-	-	X
Escambia	Pensacola Police	-	-	-	X
Hernando	Hernando County Sheriff	-	-	-	X
Hillsborough	Hillsborough County Sheriff	-	-	-	X
Hillsborough	Tampa Police	-	X	-	-
Indian River	Indian River County Sheriff	-	-	-	X
Lake	Lake County Sheriff	-	-	-	X
Lee	Lee County Sheriff	-	-	-	X
Lee	Cape Coral Police	-	-	X	-
Lee	Fort Myers Police	-	-	-	X
Leon	Leon County Sheriff	-	-	X	-
Leon	Tallahassee Police	-	X	-	-

County	Name of agency	Minimum educational level required			
		4–year degree	2–year degree	Some college*	H.S. diploma
Manatee	Manatee County Sheriff	-	-	-	X
Marion	Marion County Sheriff	-	-	-	X
Marion	Ocala Police	-	-	-	X
Martin	Martin County Sheriff	-	-	-	X
Monroe	Monroe County Sheriff	-	-	-	X
Okaloosa	Okaloosa County Sheriff	-	-	-	X
Orange	Orange County Sheriff	-	-	-	X
Orange	Orlando Police	-	-	-	X
Osceola	Osceola County Sheriff	-	-	-	X
Palm Beach	Palm Beach County Sheriff	-	X	-	-
Palm Beach	Boca Raton Police	-	X	-	-
Palm Beach	Boynton Beach Police	-	-	-	X
Palm Beach	Delray Beach Police	-	-	-	X
Palm Beach	West Palm Beach Police	-	-	X	-
Pasco	Pasco County Sheriff	-	-	-	X
Pinellas	Pinellas County Sheriff	-	-	-	X
Pinellas	Clearwater Police	-	-	X	-
Pinellas	Largo Police	X	-	-	-
Pinellas	St. Petersburg Police	-	-	X	-
Polk	Polk County Sheriff	-	-	-	X
Polk	Lakeland Police	-	-	-	X
St. John	St. John County Sheriff	-	-	-	X
St. Lucie	St. Lucie County Sheriff	-	-	-	X
St. Lucie	Fort Pierce Police	-	-	-	X

County	Name of agency	Minimum educational level required 4–year degree	2–year degree	Some college*	H.S. diploma
St. Lucie	Port St. Lucie Police	-	-	-	X
Santa Rosa	Santa Rosa County Sheriff	-	-	-	X
Sarasota	Sarasota County Sheriff	-	X	-	-
Sarasota	Sarasota Police	-	-	-	X
Seminole	Seminole County Sheriff	-	-	-	X
Volusia	Volusia County Sheriff	-	-	-	X
Volusia	Daytona Beach Police	-	-	-	X
GEORGIA					
Bibb	Bibb County Sheriff	-	-	-	X
Bibb	Macon Police	-	-	-	X
Chatham	Chatham County Police	-	-	-	X
Chatham	Savannah Police	-	-	-	X
Cherokee	Cherokee County Sheriff	-	-	-	X
Clarke	Athens-Clarke County Police	-	-	-	X
Clayton	Clayton County Police	-	-	-	X
Cobb	Cobb County Police	-	-	-	X
Cobb	Cobb County Sheriff	-	-	-	X
Cobb	Marietta Police	-	-	-	X
Columbia	Columbia County Sheriff	-	X	-	-
De Kalb	De Kalb County Police	-	-	-	X
Dougherty	Albany Police	-	-	-	X
Douglas	Douglas County Sheriff	-	-	-	X

County	Name of agency	Minimum educational level required			
		4–year degree	2–year degree	Some college*	H.S. diploma
Fulton	Fulton County Police	-	X	-	-
Fulton	Atlanta Police	-	-	-	X
Fulton	East Point Police	-	-	-	X
Gwinnett	Gwinnett County Police	-	-	-	X
Hall	Hall County Sheriff	-	-	-	X
Muscogee	Columbus Police	-	X	-	-
Richmond	Augusta-Richmond Co. Sheriff	-	-	-	X
HAWAII					
Hawaii	Hawaii County Police	-	-	-	X
Honolulu	Honolulu Police	-	-	-	X
Kauai	Kauai County Police	-	-	-	X
Maui	Maui County Police	-	-	-	X
IDAHO					
Ada	Ada County Sheriff	-	-	-	X
Ada	Boise Police	-	-	X	-
ILLINOIS					
Champaign	Champaign Police	-	-	-	X
Cook	Cook County Sheriff	-	-	-	X
Cook	Arlington Heights Police	-	-	X	-
Cook	Chicago Police	-	-	X	-
Cook	Cicero Police	-	-	-	X
Cook	Evanston Police	-	-	X	-
Cook	Oak Lawn Police	-	-	-	X

County	Name of agency	Minimum educational level required 4–year degree	2–year degree	Some college*	H.S. diploma
Cook	Oak Park Police	-	X	-	-
Cook	Schaumburg Police	-	X	-	-
Cook	Skokie Police	-	X	-	-
Du Page	Naperville Police	-	-	-	X
Kane	Aurora Police	-	-	-	X
Kane	Elgin Police	-	-	-	X
Lake	Lake County Sheriff	-	-	X	-
Lake	Waukegan Police	-	-	-	X
Macon	Decatur Police	-	X	-	-
Peoria	Peoria Police	-	-	-	X
Sangamon	Springfield Police	-	-	-	X
Will	Will County Sheriff	-	-	-	X
Will	Joliet Police	-	-	-	X
Winnebago	Winnebago County Sheriff	-	-	-	X
Winnebago	Rockford Police	-	-	-	X
INDIANA					
Allen	Fort Wayne Police	-	-	X	-
Delaware	Muncie Police	-	-	-	X
Elkhart	Elkhart Police	-	-	-	X
Howard	Kokomo Police	-	-	-	X
Lake	Lake County Sheriff	-	-	-	X
Lake	East Chicago Police	-	-	-	X
Lake	Gary Police	-	-	-	X
Lake	Hammond Police	-	-	-	X
Madison	Anderson Police	-	-	-	X

County	Name of agency	Minimum educational level required 4–year degree	2–year degree	Some college*	H.S. diploma
Marion	Marion County Sheriff	-	-	-	X
Marion	Indianapolis Police	-	-	-	X
St. Joseph	South Bend Police	-	-	-	X
Vanderburgh	Vanderburgh County Sheriff	-	-	X	-
Vanderburgh	Evansville Police	-	-	-	X
Vigo	Terre Haute Police	-	-	-	X
IOWA					
Black Hawk	Waterloo Police	-	-	-	X
Linn	Cedar Rapids Police	-	-	-	X
Polk	Des Moines Police	-	-	X	-
Scott	Davenport Police	-	-	-	X
Woodbury	Sioux City Police	-	-	X	-
KANSAS					
Douglas	Lawrence Police	-	-	-	X
Johnson	Johnson County Sheriff	-	-	-	X
Johnson	Olathe Police	-	-	-	X
Johnson	Overland Park Police	-	-	-	X
Sedgwick	Sedgwick County Sheriff	-	-	-	X
Sedgwick	Wichita Police	-	-	-	X
Shawnee	Shawnee County Sheriff	-	-	-	X
Shawnee	Topeka Police	-	-	-	X
Wyandotte	Kansas City Police	-	-	-	X

County	Name of agency	Minimum educational level required			
		4-year degree	2-year degree	Some college*	H.S. diploma
KENTUCKY					
Fayette	Lexington-Fayette County Police	-	-	-	X
Jefferson	Jefferson County Police	-	-	-	X
Jefferson	Jefferson County Sheriff	-	-	-	X
Jefferson	Louisville Police	-	-	-	X
Kenton	Covington Police	-	-	X	-
LOUISIANA					
Ascension	Ascension Parish Sheriff	-	-	-	X
Bossier	Bossier Police	-	-	-	X
Caddo	Caddo Parish Sheriff	-	-	-	X
Caddo	Shreveport Police	-	-	-	X
Calcasieu	Calcasieu Parish Sheriff	-	-	-	X
Calcasieu	Lake Charles Police	-	-	-	X
E. Baton Rouge	E. Baton Rouge Parish Sheriff	-	-	-	X
E. Baton Rouge	Baton Rouge Police	-	-	-	X
Jefferson	Kenner Police	-	-	-	X
Lafayette	Lafayette Parish Sheriff	-	-	-	X
Lafourche	Lafourche Parish Sheriff	-	-	-	X
Orleans	New Orleans Police	-	-	-	X
Ouachita	Ouachita Parish Sheriff	-	-	-	X
Ouachita	Monroe Police	-	-	-	X

County	Name of agency	Minimum educational level required 4–year degree	2–year degree	Some college*	H.S. diploma
Rapides	Alexandria Police	-	-	-	X
St. Bernard	St. Bernard Parish Sheriff	-	-	-	X
St. Charles	St. Charles Parish Sheriff	-	-	-	X
St. John the Baptist	St. John the Baptist Sheriff	-	-	-	X
St. Landry	St. Landry Parish Sheriff	-	-	-	X
St. Tammany	St. Tammany Parish Sheriff	-	-	-	X
Terrebonne	Terrebonne Parish Sheriff	-	-	-	X
MARYLAND					
Anne Arundel	Anne Arundel County Police	-	-	-	X
Anne Arundel	Annapolis Police	-	-	-	X
Baltimore(city)	Baltimore Police	-	-	-	X
Baltimore	Baltimore County Police	-	-	-	X
Charles	Charles County Sheriff	-	-	-	X
Harford	Harford County Sheriff	-	-	-	X
Howard	Howard County Police	-	-	-	X
Montgomery	Montgomery County Police	-	X	-	-
Prince George's	Prince George's County Police	-	-	-	X
MASSACHUSETTS					
Bristol	Fall River Police	-	-	-	X
Bristol	New Bedford Police	-	-	-	X

County	**Name of agency**	**Minimum educational level required** 4–year degree	2–year degree	Some college*	H.S. diploma
Essex	Lawrence Police	-	-	-	X
Essex	Lynn Police	-	-	-	X
Hampden	Chicopee Police	-	-	-	X
Hampden	Holyoke Police	-	-	-	X
Hampden	Springfield Police	-	-	-	X
Middlesex	Cambridge Police	-	-	-	X
Middlesex	Framingham Police	-	-	-	X
Middlesex	Lowell Police	-	-	-	X
Middlesex	Malden Police	-	-	-	X
Middlesex	Medford Police	-	-	-	X
Middlesex	Newton Police	-	-	-	X
Middlesex	Somerville Police	-	-	-	X
Middlesex	Waltham Police	-	-	-	X
Norfolk	Brookline Police	-	-	-	X
Norfolk	Quincy Police	-	-	-	X
Plymouth	Brockton Police	-	-	-	X
Suffolk	Boston Police	-	-	-	X
Suffolk	Revere Police	-	-	-	X
Worcester	Worcester Police	-	-	-	X
MICHIGAN					
Calhoun	Battle Creek Police	-	X	-	-
Genesee	Flint Police	-	-	-	X
Ingham	Lansing Police	-	-	X	-
Kalamazoo	Kalamazoo Police	-	-	-	X
Kent	Kent County Sheriff	-	-	-	X
Kent	Grand Rapids Police	-	-	-	X

County	Name of agency	Minimum educational level required: 4-year degree	2-year degree	Some college*	H.S. diploma
Macomb	Macomb County Sheriff	-	X	-	-
Macomb	Sterling Heights Police	-	-	-	X
Macomb	Warren Police	-	-	-	X
Oakland	Oakland County Sheriff	-	-	-	X
Oakland	Farmington Hills Police	-	-	X	-
Oakland	Pontiac Police	-	-	X	-
Oakland	Southfield Police	-	-	-	X
Oakland	Troy Police	-	X	-	-
Saginaw	Saginaw Police	-	-	-	X
Washtenaw	Washtenaw County Sheriff	-	-	-	X
Washtenaw	Ann Arbor Police	-	X	-	-
Wayne	Wayne County Sheriff	-	-	-	X
Wayne	Dearborn Police	-	X	-	-
Wayne	Detroit Police	-	-	-	X
Wayne	Livonia Police	-	X	-	-
Wayne	Taylor Police	-	-	X	-
Wayne	Westland Police	-	-	-	X
MINNESOTA					
Hennepin	Bloomington Police	-	X	-	-
Hennepin	Minneapolis Police	-	X	-	-
Ramsey	Ramsey County Sheriff	-	X	-	-
Ramsey	St. Paul Police	-	X	-	-
St. Louis	Duluth Police	-	X	-	-

County	Name of agency	Minimum educational level required			
		4–year degree	2–year degree	Some college*	H.S. diploma
MISSISSIPPI					
Forrest	Hattiesburg Police	-	-	-	X
Harrison	Biloxi Police	-	-	-	X
Harrison	Gulfport Police	-	-	-	X
Hinds	Jackson Police	-	X	-	-
Jackson	Jackson County Sheriff	-	-	-	X
MISSOURI					
Boone	Columbia Police	-	-	X	-
Buchanan	St. Joseph Police	-	-	-	X
Greene	Springfield Police	-	-	X	-
Jackson	Independence Police	-	-	-	X
Jackson	Kansas City Police	-	-	-	X
Jefferson	Jefferson County Sheriff	-	-	-	X
St. Charles	St. Charles County Sheriff	-	-	-	X
St. Louis(city)	St. Louis Police	-	-	X	-
St. Louis	St. Louis County Police	-	-	X	-
MONTANA					
Yellowstone	Billings Police	-	-	-	X
NEBRASKA					
Douglas	Omaha Police	-	-	-	X
Lancaster	Lincoln Police	-	-	-	X

County	Name of agency	Minimum educational level required 4–year degree	2–year degree	Some college*	H.S. diploma
NEVADA					
Clark	Las Vegas Metropolitan Police	-	-	-	X
Clark	North Las Vegas Police	-	-	-	X
Nye	Nye County Sheriff	-	-	-	X
Washoe	Washoe County Sheriff	-	-	-	X
Washoe	Reno Police	-	-	-	X
NEW HAMPSHIRE					
Hillsborough	Manchester Police	-	-	-	X
Hillsborough	Nashua Police	-	-	-	X
NEW JERSEY					
Atlantic	Atlantic City Police	-	-	-	X
Camden	Camden Police	-	-	-	X
Camden	Cherry Hill Police	X	-	-	-
Essex	Bloomfield Police	-	-	-	X
Essex	East Orange Police	-	-	-	X
Essex	Irvington Police	-	-	-	X
Essex	Newark Police	-	-	-	X
Essex	West Orange Police	-	-	-	X
Hudson	Hudson County Sheriff	-	-	-	X
Hudson	Bayonne Police	-	-	-	X
Hudson	Hoboken Police	-	-	-	X
Hudson	Jersey City Police	-	-	-	X
Hudson	North Bergen Police	-	-	-	X
Hudson	Union City Police	-	-	-	X

County	Name of agency	Minimum educational level required 4-year degree	2-year degree	Some college*	H.S. diploma
Hudson	West New York Police	-	-	-	X
Mercer	Hamilton Police	-	-	-	X
Mercer	Trenton Police	-	-	-	X
Middlesex	Edison Police	-	-	-	X
Middlesex	New Brunswick Police	-	-	-	X
Middlesex	Perth Amboy Police	-	-	-	X
Middlesex	Woodbridge Police	-	-	-	X
Monmouth	Long Branch Police	-	-	-	X
Morris	Parsippany Police	-	-	-	X
Ocean	Brick Township Police	-	-	-	X
Ocean	Dover Township Police	X	-	-	-
Passaic	Passaic County Sheriff	-	-	-	X
Passaic	Clifton Police	-	-	-	X
Passaic	Passaic Police	-	-	-	X
Passaic	Wayne Police	-	-	-	X
Union	Union County Sheriff	-	-	-	X
Union	Elizabeth Police	-	-	-	X
Union	Linden Police	-	-	-	X
Union	Plainfield Police	-	-	-	X
Union	Union Police	-	-	-	X
NEW MEXICO					
Bernalillo	Bernalillo County Sheriff	-	-	-	X
Bernalillo	Albuquerque Police	-	-	X	-
Dona Ana	Las Cruces Police	-	-	X	-

County	Name of agency	Minimum educational level required			
		4–year degree	2–year degree	Some college*	H.S. diploma
Santa Fe	Santa Fe Police	-	-	-	X
NEW YORK					
Albany	Albany Police	-	-	-	X
Albany	Colonie Police	-	-	X	-
Broome	Binghamton Police	-	-	-	X
Dutchess	Dutchess County Sheriff	-	-	-	X
Erie	Erie County Sheriff	-	-	X	-
Erie	Amherst Police	-	-	X	-
Erie	Buffalo Police	-	-	-	X
Erie	Cheektowaga Police	-	-	X	-
Erie	Tonawanda Police	-	-	X	-
Monroe	Monroe County Sheriff	-	-	-	X
Monroe	Rochester Police	-	-	-	X
Nassau	Nassau County Police	-	-	X	-
Nassau	Hempstead Police	-	-	-	X
New York City	New York City Police	-	-	X	-
Niagara	Niagara County Sheriff	-	X	-	-
Niagara	Niagara Falls Police	-	-	-	X
Onondaga	Onondaga County Sheriff	-	-	-	X
Onondaga	Syracuse Police	-	-	-	X
Oswego	Oswego County Sheriff	-	-	-	X
Rensselaer	Troy Police	-	-	-	X
Rockland	Clarkstown Police	-	-	X	-
Rockland	Ramapo Police	-	-	-	X
Schenectady	Schenectady Police	-	-	-	X

County	Name of agency	Minimum educational level required 4-year degree	2-year degree	Some college*	H.S. diploma
Suffolk	Suffolk County Police	-	-	-	X
Westchester	Westchester County Police	-	-	-	X
Westchester	Greenburgh Police	-	-	-	X
Westchester	Mt. Vernon Police	-	-	-	X
Westchester	New Rochelle Police	-	-	-	X
Westchester	White Plains Police	-	-	-	X
Westchester	Yonkers Police	-	-	-	X
NORTH CAROLINA					
Alamance	Burlington Police	-	-	-	X
Buncombe	Buncombe County Sheriff	-	-	-	X
Buncombe	Asheville Police	-	-	-	X
Cabarrus	Cabarrus County Sheriff	-	-	-	X
Cumberland	Cumberland County Sheriff	-	-	-	X
Cumberland	Fayetteville Police	-	-	-	X
Durham	Durham Police	-	-	-	X
Forsyth	Winston-Salem Police	-	-	-	X
Gaston	Gaston County Police	X	-	-	-
Gaston	Gastonia Police	-	-	-	X
Guilford	Guilford County Sheriff	-	-	-	X
Guilford	Greensboro Police	-	-	-	X
Guilford	High Point Police	-	-	-	X
Mecklenburg	Charlotte-Mecklenburg Police	-	-	-	X
Nash	Rocky Mount Police	-	-	-	X

County	Name of agency	Minimum educational level required			
		4-year degree	2-year degree	Some college*	H.S. diploma
New Hanover	Wilmington Police	-	-	-	X
Pitt	Greenville Police	-	-	-	X
Wake	Wake County Sheriff	-	-	-	X
Wake	Raleigh Police	-	-	-	X
OHIO					
Butler	Butler County Sheriff	-	-	-	X
Butler	Hamilton Police	-	-	-	X
Cuyahoga	Cleveland Police	-	-	-	X
Cuyahoga	Euclid Police	-	-	-	X
Franklin	Franklin County Sheriff	-	-	-	X
Franklin	Columbus Police	-	-	-	X
Hamilton	Hamilton County Sheriff	-	-	-	X
Hamilton	Cincinnati Police	-	-	-	X
Lorain	Lorain Police	-	-	-	X
Lucas	Toledo Police	-	-	-	X
Mahoning	Youngstown Police	-	-	-	X
Montgomery	Dayton Police	-	-	-	X
Stark	Canton Police	-	-	-	X
Summit	Akron Police	-	-	-	X
OKLAHOMA					
Cleveland	Norman Police	-	-	X	-
Comanche	Lawton Police	-	-	-	X
Oklahoma	Oklahoma County Sheriff	-	-	-	X

County	Name of agency	Minimum educational level required: 4-year degree	2-year degree	Some college*	H.S. diploma
Oklahoma	Oklahoma City Police	-	-	-	X
Tulsa	Tulsa County Sheriff	-	-	X	-
Tulsa	Tulsa Police	-	-	X	-
OREGON					
Clackamas	Clackamas County Sheriff	-	-	-	X
Lane	Lane County Sheriff	-	-	-	X
Lane	Eugene Police	-	-	-	X
Linn	Linn County Sheriff	-	-	-	X
Marion	Salem Police	-	-	-	X
Multnomah	Portland Police	X	-	-	-
Washington	Washington County Sheriff	-	-	-	X
PENNSYLVANIA					
Allegheny	Allegheny County Police	-	-	-	X
Allegheny	Pittsburgh Police	-	-	-	X
Berks	Reading Police	-	-	-	X
Dauphin	Harrisburg Police	-	-	-	X
Delaware	Upper Darby Township Police	-	-	-	X
Erie	Erie Police	-	-	-	X
Lancaster	Lancaster Police	-	-	-	X
Lehigh	Allentown Police	-	-	X	-
Montgomery	Lower Merion Township Police	-	-	-	X
Northampton	Bethlehem Police	-	-	X	-

County	Name of agency	Minimum educational level required 4-year degree	2-year degree	Some college*	H.S. diploma
Philadelphia	Philadelphia Police	-	-	-	X
RHODE ISLAND					
Kent	Warwick Police	-	X	-	-
Providence	Cranston Police	-	-	X	-
Providence	Pawtucket Police	-	-	-	X
Providence	Providence Police	-	-	-	X
Providence	Woonsocket Police	-	-	-	X
SOUTH CAROLINA					
Anderson	Anderson County Sheriff	-	-	-	X
Beaufort	Beaufort County Sheriff	-	-	-	X
Charleston	Charleston County Sheriff	-	-	-	X
Charleston	Charleston Police	X	-	-	-
Greenville	Greenville County Sheriff	-	-	-	X
Greenville	Greenville Police	-	-	-	X
Horry	Horry County Police	-	X	-	-
Horry	Myrtle Beach Police	-	-	-	X
Richland	Richland County Sheriff	X	-	-	-
Richland	Columbia Police	-	-	-	X
Spartanburg	Spartanburg County Sheriff	-	-	-	X
Spartanburg	Spartanburg Police	-	-	-	X

County	Name of agency	Minimum educational level required 4-year degree	2-year degree	Some college*	H.S. diploma
SOUTH DAKOTA					
Minnehaha	Sioux Falls Police	-	-	-	X
TENNESSEE					
Davidson	Nashville Metropolitan Police	-	-	X	-
Hamilton	Hamilton County Sheriff	-	-	-	X
Hamilton	Chattanooga Police	-	-	-	X
Knox	Knox County Sheriff	-	-	-	X
Knox	Knoxville Police	-	-	-	X
Madison	Jackson Police	-	-	-	X
Montgomery	Clarksville Police	-	-	-	X
Rutherford	Murfreesboro Police	-	-	-	X
Shelby	Memphis Police	-	-	X	-
Sullivan	Sullivan County Sheriff	-	-	-	X
Washington	Johnson City Police	X	-	-	-
TEXAS					
Bell	Killeen Police	-	-	X	-
Bell	Temple Police	-	-	-	X
Bexar	Bexar County Sheriff	-	-	-	X
Bexar	San Antonio Police	-	-	-	X
Cameron	Brownsville Police	-	-	-	X
Cameron	Harlingen Police	-	-	X	-
Collin	Plano Police	-	-	-	X

County	Name of agency	Minimum educational level required			
		4–year degree	2–year degree	Some college*	H.S. diploma
Dallas	Carrollton Police	-	-	X	-
Dallas	Dallas Police	-	X	-	-
Dallas	Garland Police	-	-	X	-
Dallas	Grand Prairie Police	-	X	-	-
Dallas	Irving Police	-	-	-	X
Dallas	Mesquite Police	-	-	X	-
Dallas	Richardson Police	-	-	-	X
Denton	Denton Police	-	-	-	X
Ector	Odessa Police	-	-	-	X
El Paso	El Paso Police	-	-	X	-
Fort Bend	Fort Bend County Sheriff	-	-	-	X
Gregg	Longview Police	-	-	-	X
Harris	Harris County Sheriff	-	-	-	X
Harris	Baytown Police	-	-	X	-
Harris	Houston Police	-	-	X	-
Harris	Pasadena Police	-	-	-	X
Hidalgo	Hidalgo County Sheriff	-	-	-	X
Hidalgo	McAllen Police	-	-	-	X
Jefferson	Beaumont Police	-	-	-	X
Jefferson	Port Arthur Police	-	-	-	X
Lubbock	Lubbock Police	-	-	-	X
McLennan	Waco Police	-	-	X	-
Midland	Midland Police	-	-	X	-
Nueces	Corpus Christi Police	-	-	-	X
Potter	Amarillo Police	-	-	-	X

County	Name of agency	Minimum educational level required 4-year degree	2-year degree	Some college*	H.S. diploma
Smith	Tyler Police	-	-	X	-
Tarrant	Tarrant County Sheriff	-	-	-	X
Tarrant	Arlington Police	X	-	-	-
Tarrant	Fort Worth Police	-	-	-	X
Taylor	Abilene Police	-	-	X	-
Tom Green	San Angelo Police	-	-	-	X
Travis	Travis County Sheriff	-	-	-	X
Travis	Austin Police	-	-	X	-
Webb	Laredo Police	-	-	-	X
Wichita	Wichita Falls Police	-	-	-	X
Williamson	Williamson County Sheriff	-	-	-	X
UTAH					
Salt Lake	Salt Lake City Police	-	-	-	X
Salt Lake	West Valley City Police	-	-	-	X
Weber	Ogden Police	-	-	-	X
VIRGINIA					
Alexandria(city)	Alexandria Police	-	-	-	X
Arlington	Arlington County Police	-	X	-	-
Chesapeake (city)	Chesapeake Police	-	-	-	X
Chesterfield	Chesterfield County Police	-	-	-	X
Danville(city)	Danville Police	-	-	-	X
Fairfax	Fairfax County Police	-	-	-	X

County	Name of agency	Minimum educational level required 4–year degree	2–year degree	Some college*	H.S. diploma
Hampton(city)	Hampton Police	-	-	-	X
Hanover	Hanover County Sheriff	-	-	-	X
Henrico	Henrico County Police	-	-	-	X
Loudoun	Loudoun County Sheriff	-	-	-	X
Lynchburg(city)	Lynchburg Police	-	-	-	X
Newport News (city)	Newport News Police	-	-	-	X
Norfolk(city)	Norfolk Police	-	-	-	X
Petersburg(city)	Petersburg Police	-	-	-	X
Portsmouth (city)	Portsmouth Police	-	-	-	X
Prince William	Prince William County Police	-	-	-	X
Richmond (city)	Richmond Police	-	-	-	X
Roanoke(city)	Roanoke Police	-	-	-	X
Suffolk(city)	Suffolk Police	-	-	-	X
Virginia Beach (city)	Virginia Beach Police	-	-	-	X
WASHINGTON					
Clark	Clark County Sheriff	-	-	-	X
Clark	Vancouver Police	-	-	-	X
King	King County Sheriff	-	-	-	X
King	Bellevue Police	-	X	-	-
King	Kent Police	-	-	-	X
King	Seattle Police	-	-	-	X

County	Name of agency	Minimum educational level required 4-year degree	2-year degree	Some college*	H.S. diploma
Pierce	Pierce County Sheriff	-	-	-	X
Pierce	Tacoma Police	-	X	-	-
Snohomish	Snohomish County Sheriff	-	-	-	X
Snohomish	Everett Police	-	-	-	X
Spokane	Spokane County Sheriff	-	-	-	X
Spokane	Spokane Police	-	-	-	X
Thurston	Thurston County Sheriff	-	-	-	X
Yakima	Yakima Police	-	-	-	X
WEST VIRGINIA					
Cabell	Huntington Police	-	-	-	X
Kanawha	Charleston Police	-	-	-	X
WISCONSIN					
Brown	Brown County Sheriff	-	X	-	-
Brown	Green Bay Police	-	X	-	-
Dane	Dane County Sheriff	-	-	X	-
Dane	Madison Police	-	-	X	-
Kenosha	Kenosha Police	-	-	X	-
Milwaukee	Milwaukee County Sheriff	-	-	-	X
Milwaukee	Milwaukee Police	-	-	X	-
Milwaukee	West Allis Police	-	X	-	-
Outgamie	Appleton Police	-	-	X	-
Racine	Racine Police	-	-	X	-
Waukesha	Waukesha County Sheriff	-	X	-	-

INDEX